99

DISCARD

D0206109

Environmental Ethics

CONTEMPORARY ETHICAL ISSUES

Environmental Ethics

Clare Palmer

ABC-CLIO

Santa Barbara, California
Denver, Colorado
Oxford, England

Library of Congress Cataloging-in-Publication Data

Palmer, Clare, 1967–
    Environmental ethics / Clare Palmer.
        p.   cm. -- (Contemporary ethical issues)
    Includes bibliographical references and index.
    ISBN 0-87436-840-5 (alk. paper)
    1. Environmental ethics.   I. Title.   II. Series.
    GE42.P35   1997   97-25966
    179'.1—dc21   CIP

02 01 00 99 98   10 9 8 7 6 5 4 3 2

ABC-CLIO, Inc.
130 Cremona Drive, P.O. Box 1911
Santa Barbara, California 93116-1911

This book is printed on acid-free paper ∞ .
Manufactured in the United States of America

CONTEMPORARY ETHICAL ISSUES

Contents

CONTEMPORARY ETHICAL ISSUES

Preface

Within the last decade, global climate change, species extinction, stratospheric ozone depletion, and the transportation of hazardous waste—to name but a few—have been the subjects of international negotiation and agreement. Popular concern about human relationships with other species has grown. Anxiety has been expressed about both what humans might be doing to other species (in agriculture and genetic engineering, for instance) and what other species might do to us (transmitting potentially fatal diseases, for example). Yet despite this growth in public awareness, and increasing amounts of national and international environmental policy and legislation, environmental problems do not seem to be diminishing.

There are, of course, many different reasons for this. At least one part of the explanation lies in the complex nature of environmental problems themselves, and with the ways in which they are socially, culturally, politically, and philosophically significant. Environmental issues are entwined with so many parts of our lives: what we consume and produce; where we go and how we get there; where and how we live and work; what we vote and what we value. On a global scale, environmental problems are related to political structures, international trade, wealth and poverty, science and technology, health and development. It is not surprising that environmental problems seem so difficult to understand, let alone to resolve.

But given the significance of environmental issues, it would be foolish if such difficulties discouraged us from further exploration, and there are many interesting paths that such exploration might take. Environmental ethics— the subject of this book —is just one such path. However, I would argue that it is a particularly important one. By uncovering and questioning the different ways in which the environment and other living beings can be and are valued, light may be cast on many aspects of what are understood to be environmental problems and solutions. Such study is essential if human attitudes and behavior toward the environment are to change. But environmental ethics provides no easy answers to environmental problems. It may in fact raise more questions than it resolves! Nonetheless, this book is written in the belief that the study of environmental ethics is both worthwhile in itself and an essential prerequisite for change.

Work in environmental ethics is currently chiefly located within academic institutions, often in philosophy departments. As a branch of philosophy, environmental ethics is comparatively youthful. Despite this youth, environmental ethics is located within philosophical discussions and debate about values that have lasted for thousands of years. Inevitably, then, it is often expressed in the language and vocabulary of academic philosophy. This language may on occasions seem difficult to understand and interpret. I have in this book attempted to avoid much of the technical vocabulary used by environmental ethicists. But in places this has been impossible; and elsewhere, no doubt, I have used such language without noticing it! Where possible, however, I have tried to define important unusual terms as I go along, and the glossary at the end of the book may provide further explanation.

The book is divided into ten sections. It begins with an introductory section, outlining key questions and issues in environmental ethics, and looking at some ways in which these issues have been approached in environmental ethics. Chapter 2 provides a brief chronology of the development of environmental ethics in relation to the growth of environmental concern more generally. Chapter 3 contains biographies of some of the key individuals in the development of environmental ethics, both historically and in the present. This is followed by the longest section, Chapter 4, which provides background information and an ethical investigation of 20 key environmental issues. These issues include global problems such as climate change and deforestation and key environmental concepts such as the Gaia hypothesis and sustainable development. Chapter 5 considers ethical aspects of environmental law in the United States and discusses three key court cases, while Chapter 6 examines codes of environmental ethics adopted in the public and private sphere. The concluding chapters provide information for those who wish to pursue environmental ethics further. Chapter 7 gives the names and addresses of useful institutions and organizations working in environmental ethics. Chapter 8 lists important printed resources, and Chapter 9 lists useful nonprint resources, including videos, CD-ROMs, and electronic resources. The book concludes with a glossary explaining some of the key terms used.

This book is intended both to provide an introduction to various aspects of environmental ethics and to point the way to other resources available. It may be used purely as a reference book to dip into for information about particular places, people, or issues. Alternatively, the introductory chapter and some of the other sections can be used to provide a more systematic introduction to environmental ethics, which can then be followed up by referring to other sources listed in Chapters 7 through 9. However the book is used, I hope it will provide a valuable resource for the reader.

CONTEMPORARY ETHICAL ISSUES

Acknowledgments

I would like to thank a number of people who assisted me in the research and writing of this book. My editors at ABC-CLIO, in particular Henry Rasof, supported me during most of the writing process, provided me with information relevant to the environmental law section, commented on the introduction, and helpfully extended deadlines in times of crisis. Professor J. Baird Callicott was helpful on many points of information about environmental ethics in the United States and tolerated endemic grumbling throughout the writing process; Professor Holmes Rolston provided material for the biographical sections as well as his list of useful videos on environmental ethics; and professors Paul Taylor, Andrew Brennan, and Eugene Hargrove kindly gave me detailed biographical information. Professor Hargrove and others responsible for the website at the Center for Environmental Philosophy provided up-to-date information on environmental ethics in the United States, and various members of the enviroethics e-list provided facts and references in key areas.

Closer to home, I would like to thank my colleague Dr. Peter Jones for his critical scrutiny of the introductory chapter and more generally for his academic support and his friendship over the last four years; my friends Dr. Sarah Pearce and Julie Clague, who (erratically) produced helpful references and (always) helpful amounts of alcohol; my parents for always being at the end of the telephone; and my partner Quentin Merritt for the huge amount of assistance, patience, and criticism (at least some of it constructive!) he offered while I was undertaking this project.

I would like to thank the following organizations for permission to reproduce copyright material:

The University of Edinburgh, for permission to reproduce their Environmental Policy Statement; The Coalition for Environmentally Responsible Economies (CERES) for permission to reproduce the CERES Principles; The International Chamber of Commerce for permission to reproduce the Business Charter for Sustainable Development; IBM for permission to reproduce their Corporate Policy no. 139: IBM's Environmental Policy; and Volkswagen for permission to reproduce Volkswagen's Environmental Policy.

CONTEMPORARY ETHICAL ISSUES

Chapter 1: Introduction

What Is Environmental Ethics?

On 10 February 1996, the oil tanker *Sea Empress*, carrying 140,000 tons of light crude oil, ran aground off Milford Haven in Wales. Despite several attempts to pull the ship off the rocks, she remained there for nearly a week, leaking over half of her cargo into Welsh coastal waters. Within days, oil slicks surrounded the Welsh coastline and coated the beaches of Britain's only two marine nature reserves, Lundy Island and Skomer Island.

For a short while, the quiet West Wales coast became a media mecca. Tourist accommodations lying empty for the winter were snapped up by eager television companies. Red-nosed reporters, swaddled in layers of oilskin, jostled one another for position on gale-swept, oil-acrid cliff tops. Disaster! Catastrophe! screamed the headlines. Comparisons were made with previous great oil spills: the 1967 sinking of the oil tanker *Torrey Canyon* and the foundering of the tanker *Exxon Valdez* in Prince William Sound, Alaska, in 1989.

Alongside the media, wildlife rescue teams crowded into western Wales. Makeshift hospitals for oiled birds, seals, and dolphins were set up hastily. The *Ocean Defender*, a marine wildlife rescue ship, patrolled the coasts, picking up birds and animals in distress. All oiled birds and mammals were individually cleaned, and some sent hundreds of miles to established wildlife hospitals to recuperate. Dozens of people were employed to scrape oil off the rocks and remove polluted sand. Fears were expressed that the damage to marine wildlife would be such that the area would never recover its species diversity. It was suggested that the wild heritage of future generations might have been irreparably damaged.

Although for some local people the spill meant an unexpected boost for trade during the midwinter quiet season, for others it spelled disaster. The main source of employment in this coastal area was the fishing industry; with fish contaminated with oil, and damage to other marine life vital to the food chain, local people feared that their fishing jobs might be gone forever. Others, dependent on the summer tourist trade, realized that with the specter of oil-soaked beaches and the loss of wildlife, tourism would plummet in the years to come. Demands for compensation were beginning even as the oil still leaked out of the stricken tanker. The spill threatened not only the local marine wildlife but also the livelihoods of hundreds of local people.

Though for individuals directly involved this incident was catastrophic and life-changing, for those of us remote from it, the spill may seem like just another environmental disaster, the kind of thing we hear about every day. And there are good reasons why we might think this. In many ways the foundering of the *Sea Empress* is typical of environmental disasters and of the much smaller-scale environmental problems that we all encounter throughout our lives. Looking at the incident a little more closely will therefore help cast some light on the complex characteristics of environmental problems and the ethical issues they raise.

We can begin, for instance, by looking at the causes of and responsibility for such an incident. At first glance, the cause might seem straightforward: some kind of mechanical failure on the tanker, some sort of human error, stormy weather, or some combination of these factors. This sort of reasoning might suggest that responsibility lay with the captain and crew of the ship, with the mechanics responsible for maintaining the ship, or with the company that owned the ship for not maintaining or crewing it adequately.

Though this is one way of looking at the causes of the incident and who might be responsible for it, it is certainly not the whole story. Only three months earlier another oil tanker had run aground in almost the same place, but there had been no oil spill. Why not? Because this tanker had been double hulled, a technical change that makes tankers more resistant to spillage when they run aground. The *Sea Empress*, however, had no double hull. Although a report had been published in 1992 advising the British government to make double hulls compulsory for all oil tankers in British waters, these recommendations had never become law. Had such a law been in existence, the *Sea Empress* would never have been in West Wales.

So was the British government responsible for the accident? This is one way of looking at the incident, but again it is not the whole story. After all, the oil was being transported as part of the vast global oil industry that is a central part of all industrialized economies. The lifestyles of those in industrialized economies float on a sea of oil, essential to industry, transport, and energy production. Everyone in such economies demands oil, in a variety of forms, to maintain current lifestyles (although the level of demand varies according to the wealth and activities of the individual). By creating such a demand, these individuals sanction the continued international transportation of oil at sea, and by demanding oil at the lowest prices, they encourage businesses to cut corners in safe shipping of oil products. If the question is viewed in this way, who is *not* responsible, at least indirectly, for this accident? In some sense, we all seem to be implicated.

Even so, we are not all *equally* responsible for incidents such as the grounding of the *Sea Empress*. The causes of environmental problems are complex, and responsibility for them is multilayered. This is also true of other problems discussed later in this book: the generation of wastes, climate change, species extinction. Analysis of such problems in terms of cause and responsibility is never straightforward.

The grounding of the *Sea Empress* can be used as an example of another common feature of environmental problems, whether they are single disastrous incidents, as in this case, or chronic, long-term problems: the conflicting interests of the different human beings involved.

When the ship ran aground, salvage operators spent a week trying to remove it from the rocks. Teams of people were recruited to clean the beaches. Reporters and camera crews moved into the area. All of this activity generated local employment out of season, bringing much needed income. Significant benefits and improvements in lifestyle resulted for a number of individuals. And these local people were not the only people to benefit. The pictures of oiled birds and mammals brought a flood of public interest—along with increased donations and inquiries about membership— to environmental groups. Although those working for such organizations clearly did not desire such an oil spill, the income resulting from it allowed them to plan for the future, to renew employment contracts, to run new campaigns, to advertise more widely, and to trade on the impression that their pessimistic warnings about environmental disasters were justified.

Of course, the costs of the oil spill were also significant. They were felt intensely at a local level by people employed in the fishing industry. In the short term, at least, their livelihoods collapsed; the accident had severely damaged their interests. For some local people and regular visitors, even those who only saw the spill on television, the incident was also emotionally traumatic, as a once-beautiful wild area was spoiled by deposits of oil and wildlife was harmed. The interests of people not yet born may also have been affected by the spill. Loss of wildlife diversity and even the extinction of species might result, denying future generations the opportunity to experience the same variety of life with which we are familiar.

The grounding of the *Sea Empress*, then, benefited some individuals and groups and damaged others, and the nature and degree of the damage varied over the short and the long term. In this sense the incident is like many other environmental problems. Conflicts between different human interests—employment, health, income, esthetic pleasure, lifestyle—are always tangled up in environmental problems, adding to their complexity. More generally still, in many cases the economic and social structures adopted by human societies, which lead to benefits for many members of such societies, create the circumstances that generate environmental problems. For instance, the oil industry itself contributes to the possibility of technologically sophisticated lifestyles. To this extent we benefit from its continuance, and though we may not directly benefit or suffer from particular oil spills, we gain from the persistence of circumstances that allow such spills to happen. So alongside the complex causes of, and responsibility for, environmental problems, we must also consider who benefits and who suffers from environmental problems and the circumstances that create them, both in the short and long term.

The third and final general point that can be drawn from this incident concerns another affected "party," the environment itself. The word *environment* is a difficult one to define, and I have so far used it in a rather casual way. Generally, the term refers to everything that surrounds us, the location in which we find ourselves. Clearly, if we live in an urban area our immediate environment may be apartments and shops; if we are on a ship, it is the sea and the air. However, when we talk about environmental problems, we are usually referring to problems within the natural environment, meaning those parts of our surroundings that are not human constructions. In practice, of course, boundaries here are rather difficult to draw. A conifer plantation is in some sense "natural," since trees are not human constructions, but it is also "human," since humans planted the trees and perhaps engineered the seeds. Although no distinctions here are entirely satisfactory, we could perhaps envisage the natural and the artificial as a kind of spectrum, with wilderness near to the natural end and urban landscapes near to the artificial end. This very difficulty in defining environment is compounded by the frequency with which we mean different things when we talk about environmental problems. If we return to the case of the *Sea Empress*, we can perhaps see this more clearly.

For some people, the nature of the problem after such an oil spill is restricted to the damage to the interests of human beings whose livelihoods have been seriously affected. But for many other people other issues are raised, as the scale of the cleanup operation indicates. These issues have to do with the effects of the oil spill, not on human beings but on the environment itself. Such concerns can take a number of forms. First, there are concerns for the well-being of individual animals caught in the spill who might be suffering from the effects of the oil. Second, there are concerns about particular species, especially those species that are rare; the loss of many individuals of a rare species may mean that it will become extinct. Third, there is concern for the ecology of the whole area, the delicate interactions of a wide range of species. Such a spill (by, for instance, damaging the food chain) may mean that some species will abandon the area, and the ecosystem may never be returned to its previous state.

Although any one person may be concerned about all of these issues, they are clearly different concerns, one relating to animal well-being and suffering, the second to species preservation, and the third to ecosystem health. After an oil spill, rescuing an oiled seabird might further all three aims. But there are circumstances in which these aims may conflict, and there are also ways in which most people are apparently inconsistent in their behavior.

Imagine, for instance, that the Welsh coast had recently been colonized by a species of exotic seabird that had escaped from a nearby marine wildlife park and were displacing native bird species. Suppose that, after the spill, these newcomers were suffering badly from the oil pollution. To prevent their suffering, one might feel that they should be treated like other seabirds: cleaned, treated with antibiotics, released. However, the preexisting ecosystem would be best protected by leaving these birds to die. How might one handle such a conflict of concern?

Or consider the case of a volunteer who, anxious about the suffering caused by the spill, might spend the day cleaning oiled seabirds—and return to a warming chicken soup in the evening. Is there something inconsistent about spending the day relieving the suffering of one kind of bird, while in the evening enjoying the benefits of the suffering of another kind of bird? Or is there some sense in which it is appropriate to distinguish between the treatment of wild birds and birds bred for food?

How one approaches questions such as these affects our whole understanding of what an environmental problem is. Are environmental problems primarily about the effects of environmental damage (such as pollution or species extinction) on other human beings? Or are they about causing suffering or death to individuals of other species? Or are they about preserving species or ecosystems in healthy states? Or are they a combination of all of these factors, creating a complex mesh of interrelated concerns bound into what on the surface seems to be a fairly straightforward environmental problem like the grounding of the *Sea Empress*?

This examination of the *Sea Empress* case has raised a whole series of questions. These concern the causes of, and responsibility for, environmental problems; the complicated network of benefits and costs to human beings generated by environmental problems and the circumstances that create them; and the range of concerns that may be present when we talk about environmental problems. These are the kinds of questions that are fundamental to the study of environmental ethics; we will be considering them in more detail in this book.

The Study of Environmental Ethics

As the preceding section might suggest, a one-sentence definition of environmental ethics is difficult to provide. Most fundamentally, environmental ethics examines how human beings should interact with the nonhuman world around them. What this definition means may become clearer if these two words *environment* and *ethics* are considered in turn.

Because the word *environment* can be understood very broadly, the potential scope of environmental ethics is correspondingly broad. Historically, the main focus of environmental ethics has been at the "wild" end of the natural-artificial spectrum: on wilderness areas, wild ecosystems and organisms, and biodiversity. That this has traditionally been the case, however, does not necessarily mean that it will continue to be so. More recently, some environmental ethicists have been exploring human ethical relationships with less wild aspects of the environment, such as domesticated animals and agricultural and urban landscapes.

The word *ethics* introduces the idea of *should* or *ought to*. The study of ethics is often defined as the study of how we should live and what we ought to do; what kind of behavior is right and wrong; what our moral obligations might be. Ethical statements thus differ fundamentally from descriptive ones. For instance, the statement "this bird has been poisoned by an oil spill" is descriptive. It tells us about an object (the bird), the condition of the object (poisoned), and the substance that has caused the condition (oil). However, the statements "this bird ought not to have been poisoned by an oil spill" and "it was wrong to allow an oil spill to kill this bird" are *ethical*. They make a value judgment about what has happened to the bird: it is wrong that the bird has been poisoned. (There is perhaps also the implication here that someone has acted unethically in allowing the bird to be poisoned, and that the life of the bird has a value that has been damaged or destroyed by the oil.)

If, then, we bring the words *environment* and *ethics* back together again, we can see more clearly what might be meant by saying that environmental ethics is the study of how humans should or ought to interact with the environment. Such study is far from straightforward, as the example of the *Sea Empress* illustrates. Many environmental issues that may seem at first sight to be unproblematic are, on closer scrutiny, complex and multifaceted. To deal with such complex questions, it is vital to gather relevant information and to develop some key skills.

As a first step, it is essential to acquire knowledge and understanding about any environmental question, in so far as it is available. For instance, to study the case of the *Sea Empress* in any detail, one would need to know how much oil was on the tanker, what kind of oil it was, what sort of effects current research suggests that such a quantity and type of oil might have on marine organisms, what kinds of organisms might have been present in the area, how the weather might disperse the slick, and so on. This may, in itself, be a difficult task since the knowledge surrounding many environmental questions is contested and open to diverse interpretations. In the case of the

Sea Empress, for instance, there may be profound scientific disagreement about the effects of oil on marine ecosystems both in the short term and in the long term.

Secondly, it is important to develop the skills of excavating and analyzing the different human interests, attitudes, and values that may be related to an environmental problem. There may be conflicting views about how best to approach any particular environmental problem, or even disagreement about what constitutes an environmental problem. In the case of the *Sea Empress*, someone concerned about protecting rare species may wish to target particular kinds of birds for rescue. Another individual, concerned about relieving suffering for individual organisms affected, may oppose this approach and give priority to the most seriously oiled birds. Thus, one part of the task of an environmental ethicist—we can call it a *descriptive* task—is to clarify and analyze existing human attitudes toward environmental questions and explore the values and concerns underlying them.

However, since environmental ethics concerns how humans *ought* to behave toward the environment, a more difficult task remains. This is the task of offering to others guidance or even rules about what might constitute ethically correct behavior toward the environment and what kinds of actions are right or wrong. This process is sometimes called the *prescriptive* element of ethical thinking. How to go about making ethical judgments, and what should be taken into consideration when doing so, is a topic moral philosophers have argued about for thousands of years. To do it justice here would be impossible. However, we can briefly consider some of the main approaches to ethical thinking that have been proposed over the centuries.

Natural Law and Virtues Approaches

One of the oldest ways of approaching ethics is known as the natural law tradition. This tradition takes different forms, but it is usually traced back to the writings of the Greek philosopher Aristotle (384–322 B.C.) and the later development of Aristotle's thinking by the Christian scholar Thomas Aquinas (A.D. 1225–1274). Fundamental to this approach is the belief that all things have a built-in potential, and that all things aim to fulfill this potential. An acorn, for instance, would fulfill its potential by growing, taking up nutrition, reproducing, and respiring, eventually becoming an adult oak tree. Aristotle maintains that humans also have such built-in potentials, although these go beyond growth and reproduction to include the development of intellectual and moral capacities or *virtues*. Morality, then, is an expression of human nature, and moral behavior is behavior that expresses or develops moral virtues. What might a moral virtue be? Aristotle thought that moral virtues varied among different societies but identified as virtues qualities such as courage, moderation, prudence, and justice.

There are many difficulties in accepting such views today. Aristotle, and later Aquinas, were both working within a theological framework, believing that God had created all natural objects, including human beings, with particular purposes to fulfill, and that this context provided what one might call "natural law." For many people, this theological framework is now difficult to accept, as is the idea that humans have particular capacities (such as the capacity to reproduce) that they in some sense ought to fulfill. However, the natural law tradition has been important in some modern strands of ethical thinking. It has influenced the rights approach to ethics, which will shortly be considered. Also, more recently, some philosophers have reconsidered the idea of basing ethics on the development of human virtues, though not in a theological context. Most prominent among these philosophers is Alisdair MacIntyre, whose 1986 book *After Virtue* was regarded as a groundbreaking work in modern ethics.

Deontology: Approaches Based on Rights and Duties

All of us are familiar with the idea of the rights of individual human beings. The commitment to the individual's rights to life, liberty, and the pursuit of happiness is, after all, enshrined in the heart of the Constitution of the United States. However, rights approaches to ethics are by no means straightforward.

In the Constitution, for instance, human rights are legal rights. They are rights given to people by the state, and that the state will assist them in protecting. If you try to kill me, for example, you are infringing upon my legal right to life, and assuming that the police are able to find you and that evidence against you is accepted in court, you will be imprisoned and I will be protected from you. Viewed in this way, human rights are a legal construction. More loosely, taking the United Nations Convention on Human Rights as our example, we might say that rights can be understood as an agreement between groups of humans about fundamental rules that should be respected. However—and this is where influence from the natural law tradition can be felt—many people believe that there are natural rights, rights that exist in the world independent of legal codes, in particular the right to life. From this perspective, legal codes of rights can be seen as reflecting the natural rights that already exist. It is sometimes argued that these natural rights are theologically derived and that they are God's rules, built into the universe when humans were created. But many who believe in natural rights do not offer theological reasons for them. Rather, they maintain that fundamental human rights, such as the right to life, are intuitively obvious to human beings. Indeed, they argue, this is why they attract such general acceptance. But such a view is highly controversial, and many philosophers argue that the idea of rights is created by a particular culture at a particular time and has no universal basis in intuition at all.

Another important question to ask about rights approaches, whether legal or natural, is what it means to have a right. First, rights always imply duties toward other people. Let's take the idea of right to life as an example. If you think that I have a right to life, then you are accepting that you have a duty not to take my life. And if I also think that you have a right to life, I similarly accept a duty not to take your life. Second, these duties are absolute; they cannot be revoked, no matter what the circumstances or the consequences that might result from fulfilling the duty. Ethics, then, is approached by fulfilling duties rather than by concern about the consequences of one's actions. This duty-based approach to ethics is usually called *deontological*.

As with natural law approaches, there are many different rights-based and duty-based understandings of ethics. The most important was proposed by the German philosopher Emmanuel Kant (1724–1804). Kant maintained that humans have one fundamental ethical duty: to treat other human beings as ends in themselves rather than as means to an end. This principle should never be compromised, whatever consequences might result. Kant's arguments about ethics have been hugely influential among Western philosophers and very significant in the development of thinking in environmental ethics.

Consequentialist Approaches

Although the term *consequentialist* itself may sound unfamiliar, consequentialist approaches to ethics are commonly followed in national policy making and in personal decision making. Most people make consequentialist ethical decisions many times during their lives. Consequentialism contrasts both with the natural law and virtues approaches to ethics, and with the rights- and duty-based approaches discussed above, because it is concerned with the *consequences* of actions rather than with the character or the duties of the person acting. Ethically correct behavior, according to consequentialists, is behavior designed to bring about the best consequences. So if someone makes an ethical decision based not so much on an expression of virtue, or on the fulfillment of a duty, but rather because he or she thinks that better consequences will result from this action than from any others, then that person is behaving in a broadly consequentialist manner.

Again, however, what this means in practice is far from straightforward. There is, for example, considerable disagreement about how to decide what the "best consequences" might be. What consequences should one seek to bring about when making ethical decisions? The most widely known and popular answer to this question is associated with a school of consequentialist thought called *utilitarianism*. One of the most famous utilitarian philosophers, John Stuart Mill (1806–1873), in his 1861 book *Utilitarianism*, argued that "actions are right in proportion as they tend to promote happiness, wrong as they tend to promote the reverse of happiness." Here, the best consequences are achieved by creating the greatest amount of happiness. Other consequentialists have proposed other measures of best consequences—for instance, the satisfaction of the greatest number of human preferences.

Whichever interpretation of best consequences is adopted, utilitarianism—like the other approaches to ethics we have considered—is not without difficulties. First, of course, happiness is a rather difficult thing to add up in order to decide how greatest amounts can be created. Secondly, since utilitarians recognize no absolute rules (such as the right to life), it is not difficult to imagine circumstances in which killing someone might produce the greatest happiness—for instance, killing one person in order that their bodily organs could be used to keep alive 20 individuals who would otherwise have died. Such a proposal illustrates how controversial utilitarianism can be in some circumstances.

We've now looked at three of the most important, and most debated, approaches to ethics commonly found in Western philosophy. We can shortly move on and consider how they have been, or might be, used in analyzing issues in environmental ethics. However, before doing this, there is one more central question we need to think about—the meaning and use of the term *value* in ethical thinking.

Questions about Value

All approaches to ethics rest on some understanding of value. For those who accept that humans have a right to life that should never be infringed, human life is of central value. For utilitarians like Mill, happiness is clearly the key value. But what do we mean when we talk about something or some state of affairs as being valuable? What is value? Where does it come from? Questions like these are endlessly disputed among philosophers. Here it will only be possible to make a few key distinctions, but they are absolutely crucial in understanding problems and controversies in ethics—especially in environmental ethics, where questions about value are hotly disputed.

The first helpful distinction is one commonly drawn between instrumental and noninstrumental value. Instrumental value is value given to something because of its usefulness to us. Air, for instance, is useful to us because it keeps us alive. We thus value it instrumentally; it is not valuable in itself but because it helps us to achieve another goal—remaining alive. We value air as a means to an end. However, this does not seem to be the case with all kinds of value—for instance, staying alive. We do not value our lives for any reason beyond themselves; we do not (usually) regard preserving our lives as a means to an end, but rather as an end in itself. Value of this sort is noninstrumental value (sometimes also called *intrinsic* value, a usage that will be adopted in this book).

This distinction is of particular importance in environmental ethics, where questions about value in the natural and living world are central. Supposing we take Yosemite National Park as an example. This park is widely thought to be a valuable landscape. But why is it valuable? We can cite a number of ways in which Yosemite might be instrumentally valuable to human beings. We might value it as a place to pursue leisure activities such as hiking, or as a magnificent area to paint or photograph. We might value it esthetically as stunning mountain scenery, as a place to retreat from the stresses of urban living, as a place with strong spiritual or ancestral connections. All of these are instrumental values. But does Yosemite have noninstrumental or intrinsic value? Can we value Yosemite not because of its usefulness but because it is a valuable thing, in the same way that we think human lives are valuable things? Many environmental ethicists have argued that landscapes, other organisms, and ecosystems are of intrinsic value.

The discussion of intrinsic value inevitably raises a second question about where such value comes from. Is it created by human beings, or is it something that already exists in the world, something that human beings recognize rather than bring into being?

This question, too, is a subject of great debate among philosophers, sometimes called the dispute between value *subjectivists* and value *objectivists*. Value subjectivists think that intrinsic value is something humans create and attach to their own lives and the lives of other people, to particular states of affairs (such as pleasure or the avoidance of suffering), and perhaps to landscapes or animals as suggested above. Value objectivists, on the other hand, think that intrinsic value is not something humans create but something that is built into the world around them. They might argue that in valuing Yosemite National Park they are not *creating* value but rather *recognizing* value already present, value that would continue to exist even if there were no human beings left in the world to value it.

There are clearly difficulties with holding to such an objectivist view. What kind of thing or quality is value? Is it a quality possessed by objects or individuals, rather like their color? What if people disagree (as they do) over what objects or individuals might have this quality? Who is to decide who is right? For those who believe in God, dealing with these kinds of questions is less problematic. If God created the world and (as in the Judeo-Christian tradition at least) saw that it was good, then humans can recognize and value God's work in the world. This theological understanding of value is sometimes found among those who take an approach to ethics based on natural law. But for those who do not accept that ethics has a theological basis, it is more difficult to explain what objective value might be and where it might come from. For this reason, many (although not all) of those working in environmental ethics are value subjectivists; they think that environmental values are constructed by humans.

Analyzing Issues in Environmental Ethics

The three traditional approaches to ethics discussed above, namely natural law and virtues approaches, rights and duties approaches, and consequentialist approaches, do not explicitly address the ways we might analyze issues in environmental ethics. These traditions developed when environmental issues—as we are familiar with them today—were not problems that philosophers (or anyone else) were grappling with. Philosophers were primarily concerned with how human beings related to each other (and in religious traditions of ethics, with God), not with how humans might relate to their environments.

The increasing significance of environmental issues, then, raised difficulties for those working within traditional approaches to ethics. Three different kinds of response to the difficulties emerged. The first maintained that the ethical questions raised by environmental issues were part of the traditional range of ethical concerns; that is, environmental problems were fundamentally about problems in relationships between human beings. The second argued that traditional approaches to ethics needed extending and reworking to accommodate questions raised by environmental issues. The third came from outside traditional approaches to ethics altogether and developed new ways of analyzing environmental issues. So let's look at these different kinds of response in more detail.

Applying Traditional Ethical Analyses to Environmental Issues

In much popular discussion of environmental issues, as well as most public policy making, a traditional, human-focused approach to ethics is adopted. The environment is regarded as something that has instrumental value. Elements of the environment provide human beings with valuable resources for food, warmth, industry, and leisure. Human actions in this useful environment are problematic when they have damaging implications for humans (by causing pollution, for example, or by using up resources). An incident such as the oil spill from the *Sea Empress* would be viewed ethically in terms of those humans responsible for, and affected by, the incident.

Of course, exactly how such an incident is conceived ethically would depend on which of the traditional approaches to ethics is adopted. A utilitarian, for instance, would examine the good and bad consequences of the oil spill for human happiness and recommend actions that would increase the good consequences and reduce the bad. A deontologist, however, concerned with human rights and human duties, would analyze the incident in terms of the dereliction of duty involved (perhaps of the crew, or the port authorities, or the government, or all consumers of oil) and the infringement of rights, for instance, the right to pursue a legal and chosen trade such as fishing.

However, what is common to all these traditional ethical approaches is the emphasis on human aspects of the spill, in terms of human character, intention, duty, and suffering. For this reason, such approaches to ethics are often called *anthropocentric* (meaning human-centered): it is human beings who are the center of ethical concern.

Extending Traditional Analyses

Although such human-centered approaches have characterized much debate and public policy making, environmental ethicists have developed ways of extending traditional approaches to take into account the new questions raised by environmental issues. Many different kinds of revisions and extensions have been proposed, and I shall look here at just a few of the most significant.

Utilitarianism is the tradition that most obviously lends itself to some form of redefinition. Even in its best-known form—the ethical obligation to create greatest happiness—utilitarianism could be thought of as including animals. Certainly, Jeremy Bentham (1748–1832), the founder of utilitarianism, thought so. In his *Introduction to the Principles of Morals and Legislation* (1789), he wrote of animals, "The question is not Can they *reason?* nor Can they *talk?* but Can they *suffer?*" This understanding of utilitarianism as including animals was largely neglected until the utilitarian philosopher Peter Singer published his book *Animal Liberation* in the early 1970s. This book is now viewed as one of the key works in the animal liberation and environmental movements. In it, Singer argued that because animals are capable of feeling pain and pleasure, they should be taken into account when we make ethical decisions. Inflicting suffering on animals is wrong, unless by doing so greater suffering is averted. Causing suffering for reasons of taste (as in meat-eating) or appearance (as in the testing of cosmetics on animals) is thus ethically unacceptable.

The significance of Singer's arguments for the development of environmental ethics is that organisms other than humans are directly taken into account in moral decision making. They are not being valued as resources, for their usefulness to human beings, but rather for what they are in themselves. Indeed, they may be valued in ways that may cause inconvenience or even difficulty for human beings. Animals are thus seen as having intrinsic value. The focus of such an approach is, then, no longer anthropocentric. Viewed from this perspective, the oil spill from the *Sea Empress* gives rise to a whole new range of ethical concerns. The leaked oil clearly caused suffering to marine mammals and birds; for this reason, Singer's position suggests, suffering should be minimized by cleaning up the oil where possible, by treating birds and marine mammals that have come into contact with it, and by taking steps to prevent a recurrence.

Clearly, then, Singer's utilitarianism does have important implications for the analysis of environmental issues. But many environmental ethicists felt that Singer's approach did not go far enough. The natural environment is, for Singer, still a "backdrop" to those organisms that can feel pleasure and pain. Other elements of the environment—plants, ecosystems, many species—are still only valued as long as they are useful. To some thinkers, this was not an adequate approach to environmental ethics. Thus, some philosophers, while still adopting a consequentialist approach, proposed a change in direction. They suggested that rather than focusing on pain and pleasure as the key to analyzing environmental problems, well-being should be the focus. Well-being is a concept, they argued, that could be applied to all living organisms, since all can be in states that are better or worse for them. It is not necessary, they argued, to be able to feel pain to have one's well-being damaged. Such a criterion would suggest that all living organisms—plants, insects, and bacteria—should be directly taken into account when analyzing ethical problems. Indeed, some environmental ethicists went further and maintained that the idea of well-being applies not only to individual organisms but also to groups of organisms, such as species and ecosystems. We have no difficulty making sense of the idea that ecosystems can be healthy or unhealthy, they argue, and that their well-being can be damaged or promoted. Thus, we need to take these groups, or "collectives," into account along with individual organisms when making ethical decisions about the environment.

Developing a utilitarian framework in this inclusive way produces a whole new range of environmental ethical concerns. The consequences of an oil spill such as that from the *Sea Empress* now seem much more extensive. The human and animal concerns still exist; the spill impairs their well-being. But the well-being of countless other organisms has also been affected, including plants and plankton. In addition, the well-being or health of the whole ecosystem has been damaged. The ethical consequences of the spill now seem far more widespread and thus more difficult to address or to compensate for.

But an emphasis on ecological collectives also raises other questions for environmental ethics. What happens in instances where the well-being of individual organisms conflicts with that of the ecological collective (for instance, where, as in the earlier example, exotic birds that are harmful to a native ecosystem are suffering)? Should the individuals' suffering outweigh the well-being of the ecosystem? Or should the well-being of the ecosystem outweigh the suffering of the individuals? Different environmental ethicists have proposed different answers to this dilemma, and the disagreement has led to a dispute between animal liberationists (who put the well-being of individuals first) and some environmental ethicists (who put the well-being of ecological collectives first). An important collection of essays published in 1992 and edited by Eugene Hargrove called *The Animal Rights/Environmental Ethics Debate* explores this debate between individualists and collectivists much more fully.

Alongside these developments and disagreements within a broadly consequentialist tradition, similar developments have occurred in rights and duty approaches to ethics. In 1983, partly in response to Singer's *Animal Liberation*, the American philosopher Tom Regan published a book called *The Case for Animal Rights*. In this book Regan, who accepts that humans have natural rights, argued that animals (or to be more precise, adult mammals) also possess natural rights, in particular the right to life. These rights, he argued, like human rights, are inviolable.

Regan's position, like Singer's, clearly has implications for any analysis of environmental issues. Neither his proposal nor Singer's is anthropocentric. But again, some philosophers thought Regan did not go far enough in including the environment in our ethical concerns. A range of attempts to develop more environmentally inclusive ethics within the rights and duty approaches has since been made. Perhaps the most important of these was proposed by the American philosopher Paul Taylor in his 1986 book *Respect for Nature*. Although Taylor does not himself use rights language, he does develop a detailed theory of human duties toward living organisms. Drawing on the work of Aristotle and Kant, Taylor argues that all living things have their own purpose; they are, as he says, "ends-in-themselves." They are not instruments for us to use; they have value in themselves, and they have this value equally. Thus, he maintains, humans have duties to respect all living organisms equally. These duties include refraining from harm and interference (except in particular circumstances, such as to save life). Taylor, however, does not extend his position to include ecosystems or species; only individual organisms, he maintains, are ends-in-themselves, and only toward them do humans have duties.

Taylor's position creates new dimensions to the ethical analysis of environmental issues. The *Sea Empress* oil spill would be viewed by Taylor as a dereliction of duty on a number of counts. As well as neglecting human duties toward other humans, the spill is also a betrayal of human duties not to harm or interfere with other living organisms. Thus, like some of the reformulations of consequentialism considered above, Taylor's approach takes all living organisms directly into account in ethical decision making.

It is clear, then, that much work in environmental ethics has attempted to reformulate traditional approaches to ethics in order to address the new questions raised by environmental issues. Many more similar developments could be considered—in the virtues tradition as well as in specifically utilitarian or more broadly consequentialist and rights traditions. But alongside such specifically ethical approaches to the environment, broader philosophical and political movements have arisen, contributing important insights to the development of environmental ethics. Most important among these are deep ecology, social ecology, and ecofeminism.

Deep Ecology, Social Ecology, and Ecofeminism

It is important to note that deep ecology, social ecology, and ecofeminism are *umbrella* terms. Each may encompass a variety of differing positions. It is also helpful to be aware that ethics is only part of what deep ecology, social ecology, and ecofeminism are about. They are, in effect, much broader philosophical-political movements from which particular interpretations of ethics flow.

Deep ecology was a term first used in print by the Norwegian philosopher Arne Naess in 1973. Naess argued that the environmental movement had two key strands, which he called the "shallow" and the "deep." The shallow movement, he maintained, was primarily concerned with human welfare and with issues such as the exhaustion of natural resources. In contrast, the deep environmental movement (with which Naess identified himself) was concerned with fundamental philosophical questions about the ways in which humans relate to their environment. In particular, deep ecology incorporated insights from modern physics and ecology into human understanding of the natural world. Much Western philosophy, Naess argued, relies on an outdated view of the world, in which human beings are believed to be separate from one another and from the natural world. But recent work in physics and ecology, Naess argued, does not support such an understanding of the world. The new science, he maintained, understands humans not as isolated, separate objects but rather as interconnected with each other and constantly in relationship with everything around them—part of the flow of energy, the web of life. From this perspective, the first priority in analyzing environmental issues must be transforming one's fundamental way of looking at the world, to develop what Naess calls a more "holistic" outlook.

Though deep ecology has developed and taken many forms since 1973, this plea for a profound change in the way many westerners think about the world is still at its heart. Not surprisingly, deep ecologists argue that such a change in worldview entails a corresponding change in values and ultimately a change in our environmental ethic. In the early days of the movement, deep ecologists argued that all living beings should be regarded as having equal value (not dissimilar to the position of Paul Taylor). More recently, some have rejected this approach altogether. One prominent deep ecologist, Warwick Fox, in his book *Toward a Transpersonal Ecology* (1991), has not only rejected particular approaches to ethics but the idea of ethics altogether. He maintains that deep ecology "renders ethics superfluous." He argues that if we accept the deep ecology worldview—that in some profound sense we are not separate from the world around us—then it is difficult to make sense of the idea of having duties toward others, because there can be no real others. No one and no thing can be fully separated from ourselves; we are too closely connected. Our selves extend into the world around us; our actions in the world are thus really actions toward ourselves; other people's actions toward the environment are actions toward us.

If we analyze the *Sea Empress* spill from this perspective, a very different picture emerges from that produced by other positions we have considered. The spill is a violation of ourselves; a cause of grief and suffering. Harm to the natural world is harm to ourselves, protection of the natural world is self-defense. The personal nature of this offense might then lead us to offer assistance in the cleanup, to campaign against the possibility of such accidents happening again, to mourn the loss of an unblemished part of ourselves.

Even this brief summary is sufficient to indicate that deep ecology offers a very different way of analyzing environmental issues, and thus a different approach to environmental ethics, than the more traditional approaches previously outlined. Social ecology, the second school of thought to be considered here, also offers a rather different analysis, and one that is firmly opposed to the deep ecology approach.

Social ecology was founded by Murray Bookchin, an American philosopher and political thinker, during the 1970s. He was influenced by the writings of one of the architects of communism, Karl Marx, and also by anarchist political traditions that oppose the idea of human societies being ruled by governing elites. Bookchin argues that the roots of environmental problems lie in human relations to one another rather than (as deep ecologists suggest) in human misunderstanding of their connection with the natural world. Many human societies, he maintains, are organized into hierarchical layers of dominance and oppression, where some classes of people and some kinds of human qualities or abilities are regarded as superior to others—for instance, men are thought superior to women, intellectual skills superior to physical skills. The idea that the natural world is inferior to human beings and there to be exploited and abused stems from, and is an expression of, the hierarchical nature of human relationships with one another.

Thus, it follows logically, Bookchin argues, that in order to resolve environmental problems, humans' relationships with one another must first be changed. He rejects the idea that humans should live in societies of hierarchical relationships in which some individuals are dominant over others. Rather, he advocates social equality and freedom, giving examples from preliterate societies that he believes demonstrate these characteristics. Once there is equality and an end to dominance in human societies, the end of dominance of nature will follow. So changing attitudes to the environment can only occur once a hierarchical view of human society has been rejected.

Bookchin's approach, then, contrasts with that of most environmental ethicists and all deep ecologists. His insistence that human relationships with one another must change before attitudes toward the natural world can change has, in fact, been opposed by many environmental ethicists and deep ecologists. They argue that his position, like traditional philosophical approaches, is anthropocentric. Certainly, a social ecological analysis of environmental issues would look quite different from those we have already considered, as we can see by returning to the problem of the oil spill from the *Sea Empress*. Social ecologists might argue that this incident illustrates how the privileged members of industrialized societies (such as the managers and executives of oil companies and members of governments who act as consultants for them) pursue their own gain at the expense of poorer members of society (such as members of the fishing community) and the environment. But to do something about such incidents requires a transformation of the nature of human society. In a human society without such oppressive hierarchies, a social ecologist would argue, such environmental destruction would not occur.

The final nontraditional way of analyzing environmental issues that I want to consider here is ecofeminism. The term *ecofeminism* was first used in

1974 by Francoise D'Eaubonne. Since then ecofeminism has become a large but diverse movement, encompassing a wide range of perspectives from within the feminist movement and the environmental movement. What all ecofeminists have in common is the view that there is a link between the domination of nature and the domination of women. In some respects, this view is similar to that of social ecology—human society is organized in hierarchical ways, which leads to the oppression of women and the natural world. However, ecofeminists do not argue, like social ecologists, that human society must be changed before changes in attitudes toward the natural world can occur. Rather, they argue that oppression of women and the natural world are twin oppressions that must be addressed together.

There are several different kinds of ecofeminists. Some argue that women are somehow closer to nature than men, while others reject this view. Some come from within a broadly liberal political tradition, while others have a socialist background. We cannot consider all these different understandings here, but it is helpful to look briefly at ecofeminist approaches to environmental ethics. All the approaches to ethics we have considered so far have proposed abstract, universal ethical principles, which were then applied to a range of different circumstances. For example, Tom Regan maintains that all adult mammals have a right to life. Rights are clearly not material things; they are abstract and conceptual. Furthermore, in Regan's arguments at least, they are also universal—they apply to all adult mammals everywhere, regardless of the reigning culture or religion, or the time of their existence.

Ecofeminist environmental ethicists have tended to reject this abstract and universalist approach to ethical thinking, which takes no account of different cultural perspectives and different kinds of relationships between humans, animals, and the natural world. Rather, they have argued for much more local, context-specific approaches to environmental ethics based on specific human relationships to particular environments. As Karen Warren, an important ecofeminist, argues in her essay *The Power and Promise of Ecological Feminism* (1992) for ecofeminists, ethics should grow out of key relationships in the lives of those concerned rather than resulting from the application of abstract, universal principles to particular situations. Clearly, this approach might produce very different ethical beliefs and behavior from the abstract, universal principles of environmental ethics we considered earlier. It also suggests that the "wrong" involved in an incident such as the oil spill from the *Sea Empress* will vary according to the perspective of the individuals concerned. For instance, for those who lived in local communities, who had over many years built up a relationship of care and affection toward the local environment, the spill might be regarded as a harm in the way one might be harmed by an assault on a friend. One might respond to this by attempting to mitigate the effects of the spill (as one might nurse a friend) and by bringing those responsible to justice (as one might try to ensure those who assaulted a friend were arrested and tried). At the root of ecofeminist ethical approaches to the environment, then, is an ethics of relationship and care, the manifestation of which will vary from place to place and relationship to relationship.

This section has indicated, then, that there are many different ways of ethically analyzing environmental issues. Of course, these different approaches lead to disagreement and controversy, both in regard to the general principles that should be followed (if any) in environmental ethics and in the response to particular environmental issues and problems.

The Major Issues Today

Today, environmental ethicists face a wide range of questions. At the most fundamental level, they are concerned with debates about human nature—how to understand and interpret the nonhuman natural world, how the nonhuman and human worlds are related, and what responsibilities this relationship might generate for humans. They also explore current environmental issues, asking why particular issues are of concern and what (if anything) is important about them. Indeed, it is quite possible that an environmental ethicist might conclude that an issue of great popular concern is not really of much ethical importance. For instance, some ethicists argue that the whole concern over species extinction is a red herring. The idea of a species, they maintain, is a human construction imposed on the natural world. The loss of such a construction is not important; a species is not the kind of thing that is valuable (although individual members of the species, or the ecosystem in which the species is located, may be valuable). From this perspective, the whole public outcry over the loss of species is mistaken; ethical concern is being expressed about something that is not really valuable.

More commonly, however, environmental ethicists work within the framework of current environmental issues, such as wilderness protection, seeking to explore and develop ethical perspectives on them. Indeed, the debate about wilderness is central to work in environmental ethics, focusing in particular on the meaning and significance of the "wild" and how wilderness areas should be valued. Are there parts of the earth that are pure and undefiled by people? What does holding such a position imply about the place of humans in the natural world? Are places where humans do not and have not lived more valuable than other kinds of places? If so, are rural, agricultural, and urban landscapes less valuable? The debate about wilderness is a good example of an issue where highly theoretical questions about nature and value have substantial practical and policy implications. If, for instance, wildernesses are to be valued as places where there are no people, does this imply that indigenous peoples should be moved out of, or prevented from entering, wilderness (a policy that has been adopted in some parts of the world)? Does this itself raise new ethical questions?

Other issues, discussed in more detail in Chapter 4, also bring together fundamental questions about nature and value with more practical matters of policy and practice. For instance, modern transportation systems, especially private motor vehicles, provide huge benefits to some human beings, yet the environmental costs in terms of pollution (by exhaust emissions or oil spills such as the *Sea Empress*), loss of habitat, environmental harm, and wildlife deaths from road accidents are very high. How do we make ethical judgments about such costs and benefits? How does the esthetic attraction of wild areas and the lives of the wild organisms who live there rank alongside the human preference to travel in speed and comfort? Or is it inappropriate to think of these effects in terms of cost and benefit at all? Are there some "wild" areas, for instance (like Yosemite National Park), that should not be destroyed to create a road, however much benefit to humans a road might bring? What arguments might be used to support such a position? Issues such as tourism, energy and waste production, deforestation, resource use, pollution, and climate change also raise fundamental questions about human interactions with the environment and about the nature and importance of environmental values.

Environmental Ethics and the Future

Interest in environmental ethics, both as a matter for general discussion and as an academic subject taught in colleges and universities, has grown since environmental questions were first widely discussed in the late 1960s. Whether this growth in interest will continue, reach a plateau, or decline in the future depends on a range of factors. In part, the political climate, both in the United States and internationally, will determine how environmental issues are regarded. The so-called backlash against the environmental movement and moves to repeal environmental legislation suggest that environmental issues will be less politically significant in the future. However, another major factor, of course, is what actually happens to local, national, and global environments. Environmental disasters (such as the atomic accident at Three Mile Island) or increased drought and storm due to climate change could raise the profile of environmental issues and pose important questions for environmental ethics.

The direction that environmental ethics might take in the future is as uncertain as the amount of interest it will generate. It is unlikely that ethicists will reach any consensus about methods of approaching ethics or the nature of intrinsic value. It is possible that ethicists will develop more local, contextual approaches to ethics (like those put forward by some ecofeminist ethicists) rather than attempting to create universal, overarching ethical systems that apply to everyone. More certain, however, are the issues with which environmental ethicists will become increasingly concerned.

First, it is likely that environmental ethicists will continue to be concerned with the currently significant issues, such as the meaning, function, and permissible uses of wilderness in the United States and throughout the world. Discussion of this issue will continue to include not only how wilderness questions should be addressed ethically but also what sorts of actions can be appropriately taken to express moral disagreement with policy decisions— demonstrations, peaceful illegal occupation of threatened areas, or violent action. Such issues, which arose during the civil rights movement in the 1960s, are current in the environmental movement and will no doubt continue to be tackled by environmental ethicists in the future.

Complementing wilderness protection as an issue of ethical concern will be ethical interest in other kinds of environments, for a number of reasons. Wildernesses are declining both in size and number. Urbanization and population are expanding. Most people in the world rarely or never enter wild areas, as they live and work in urban or rural agricultural areas. Environmental ethicists living in these populated areas are beginning to argue that *their* environments raise ethical questions, too, although the questions raised may be different ones. It seems likely that the body of work exploring urban and agricultural environmental ethical questions will grow.

Finally, it seems likely that environmental ethicists will expand their work in the area of technological change, tracking—and even attempting to predict—developments in technology and their possible environmental and ethical implications. Increasingly, genetic engineering will be a topic of discussion and debate, due both to the theoretical philosophical questions raised about what is "natural" and the more practical environmental and animal welfare implications of much work in biotechnology.

However, predicting what the future holds for environmental ethics is a precarious business because of the extent of political, social, technological, and environmental unknowns. What seems certain is that as long as there are environmental problems there will be ethicists thinking—and disagreeing— about the questions they raise.

CONTEMPORARY ETHICAL ISSUES

Chapter 2: Chronology

1650   René Descartes, French philosopher, argues in his *Meditations* that the nonhuman world, including animals, is without soul, unfeeling, and mechanical.

1690   John Locke, English political philosopher, outlines his influential theory of property in *Two Treatises of Government*. He argues that individuals have a right to own land as property if they have mixed their labor with it (cultivated or developed it).

1789   English political and moral philosopher Jeremy Bentham in *An Introduction to the Principles of Morals and Legislation* maintains that animal suffering should be taken into account when moral decisions are made. He argues that "The question is not Can they *reason*? nor Can they *talk*? but Can they *suffer*?"

1845   The American naturalist and philosopher Henry David Thoreau goes into

1845
cont.   wilderness retreat at Walden Pond, Massachusetts. His journals from this time form the basis of his influential work *Walden*.

1859   Publication of Charles Darwin's *On the Origin of Species*.

1864   Publication of George Marsh's *Man and Nature*, the first comprehensive description in English of human destruction of the environment.

1872   Creation of Yellowstone National Park (the first U.S. National Park) as a "pleasuring ground" for Americans.

1890   Creation of Yosemite, Sequoia, and General Grant National Parks.

1891   Publication of the English moralist Henry Salt's important work *Animal Rights in Relation to Social Progress*, in which he maintains that humans and animals have a "brotherhood."

1892   Founding of the Sierra Club on 4 June. Naturalist, writer, and philosopher John Muir is elected as its first president, and there are 182 charter members.

1908   The Sierra Club, headed by John Muir, launches an ultimately unsuccessful attempt to protect a wilderness area at Hetch Hetchy in California from development as a dam to provide water for San Francisco. The failure of this campaign in 1913, when Hetch Hetchy is finally granted to San Francisco, ultimately leads to Muir's death in 1914.

1914   Death of the last passenger pigeon—an extinction that gains a high public profile.

1933   The German doctor, philosopher, and musician Albert Schweitzer publishes *Civilisation and Ethics*. In this work he puts forward his famous principle of reverence for life, later to become significant in the development of environmental ethics.

1935   The biologist Charles Tanzley first uses the word *ecosystem*.

1948   Writer and forester Aldo Leopold dies in a fire.

1949   Posthumous publication of a collection of Leopold's essays in *A Sand County Almanac*. In these essays, Leopold argues that a new land ethic is required to preserve the "stability, integrity, and

1949
cont.
beauty of the biotic community," of which human beings are just "plain members and citizens." This book is to become very influential in American environmental ethics.

1962
Publication of Rachel Carson's *Silent Spring*, which examines the effects of pesticide and herbicide use on wildlife and human health. This book marks the beginning of the modern environmental movement.

1964
U.S. Congress passes the first Wilderness Act, defining wilderness as a place where "the earth and its community of life are untrammeled by man, where man himself is a visitor who does not remain." This is the first wilderness legislation in the world. Congress also establishes the Land and Water Conservation Fund.

Stewart Udall, secretary of the interior under John F. Kennedy, publishes *The Quiet Crisis*, which feeds the growth of the environmental movement.

1966
U.S. Congress passes an Animal Welfare Act, mainly concerned with the welfare of domesticated animals and animals kept for experimental purposes.

1967
History professor Lynn White publishes an article called "The Historic Roots of our Environmental Crisis" in the journal *Science*. He argues that Christianity is in part to blame for the environmental crisis because in its Western form it has desanctified and dominated nature.

1969
Foundation of the environmental activist organization Greenpeace.

1970
President Nixon, in his State of the Union address, defines the great question of the 1970s as how to secure an "unpolluted environment" as the "birthright of every American."

First Earth Day held.

Major environmental legislation, the National Environmental Policy Act (NEPA), is passed, and the Environmental Protection Agency (EPA) is established.

1971
First major conference on environmental philosophy is held at the University of Georgia. (Proceedings are published in 1974 as *Philosophy and Environmental Crisis*.)

1972   The Marine Mammal Protection Act is passed.

Law professor Christopher Stone publishes a paper called "Should Trees Have Standing?" in the *Southern Californian Law Review*, in which he argues that legal rights should be extended to natural objects.

1973   A rigorous Endangered Species Act is passed in the United States.

The philosopher Peter Singer publishes an article called "Animal Liberation" in the *New York Review of Books* (published in more detailed form in 1975 as a systematic philosophical work in the utilitarian philosophical tradition). This work is important in the rapid expansion of international movements for animal liberation. Singer argues that any organism that is capable of feeling suffering and enjoyment has interests, and that all interests should be given equal moral consideration when making moral decisions.

Arne Naess's article "The Shallow and the Deep, Long-Range Ecology Movement: A Summary" is published in the international philosophical journal *Inquiry*. This distinction between "shallow" and "deep" approaches to environmental questions ultimately gives birth to the deep ecology movement.

1974   Christopher Stone's ideas in "Should Trees Have Standing?" are attacked by philosopher Mark Sagoff in the *Yale Law Journal*. He argues that since we cannot know what natural objects might want, it does not make sense to extend rights to them, and cultural and esthetic considerations are better reasons to protect wild areas.

John Passmore, an Australian philosopher, publishes a book-length philosophical work on environmental problems called *Man's Responsibility for Nature: Ecological Problems and Western Traditions*. This book rejects deep approaches to environmental concerns in favor of some more traditional philosophical approaches.

Francois D'Eaubonne first uses the term *ecofeminism* in her book *Le Feminisme ou la Morte*.

1975   Edward Abbey publishes *The Monkey Wrench Gang*, a novel about a group of environmental saboteurs who attack equipment and constructions that are environmentally damaging. The book inspires a more active approach to protests about environmental destruction.

1975
cont.
Rosemary Radford Reuther publishes one of the first works to link feminism and environmentalism, *New Woman, New Earth*. She argues that the environmental movement and the women's movement should unite to put forward a new value system.

1978
Publication of the biologist David Ehrenfeld's book *The Arrogance of Humanism*, in which he argues that all species have a right to continued existence, including disease species like the smallpox virus.

First edition of the journal *Environmental Ethics* is published, edited by the philosopher Eugene Hargrove.

1979
The English scientist James Lovelock puts forward his Gaia hypothesis for the first time in *Gaia: A New Look at Life on Earth*.

Nuclear accident at Three Mile Island in Pennsylvania leads to increased doubts about the safety of nuclear power as an energy resource.

1980
Formation of Earth First!, a radical environmental group committed to the principle "No Compromise in Defense of Mother Earth," by environmental activist Dave Foreman. Edward Abbey's *The Monkey Wrench Gang* influences the group's philosophy.

The philosopher J. Baird Callicott publishes "Animal Liberation: A Triangular Affair" in the journal *Environmental Ethics*. He argues that the environmental movement and the animal liberation movement are not allies (as was popularly assumed) but are based on fundamentally different principles. The animal liberation movement, he argues, is based on the welfare of individual members of animal species, while the environmental movement is concerned for the well-being of the biotic community as a whole. These understandings of what is ethically significant, he maintains, are radically different and result in very different approaches to environmental and animal welfare policy.

Three Australian philosophers, Don Mannison, Michael McRobbie, and Richard Routley (later Sylvan), publish a collection of essays called *Environmental Philosophy*. Some of the essays, in particular "Human Chauvinism and Environmental Ethics" by Richard and Val Routley (later Plumwood), become important in environmental ethics.

Congress passes the Alaska National Interest Lands Conservation Act. This act designates 104 million acres of land in Alaska as parks, wildlife reserves, and wilderness.

1982   Social ecologist Murray Bookchin publishes *The Ecology of Freedom*, a key book in the development of the social ecology movement.

1983   An important systematic work in the theory of animal rights, *The Case for Animal Rights*, is published by the philosopher Tom Regan. He argues that all adult mammals, including human beings, possess natural rights.

1985   The deep ecologists Bill Devall and George Sessions publish *Deep Ecology: Living as If Nature Mattered*, a volume that is sometimes called a manifesto for the deep ecology movement. The book includes what is called the Deep Ecology Platform, a series of principles fundamental to the understanding of deep ecology, compiled in discussion with Arne Naess.

1986   A Russian nuclear reactor at Chernobyl in the Ukraine explodes, depositing radioactive fallout over Europe and causing long-term damage to human health and the natural environment. This disaster raised the profile of environmental issues internationally.

Respect for Nature, an important systematic work in environmental ethics, is published by the philosopher Paul Taylor. Taylor argues that all individual living organisms are equally valuable. This argument provides grounds for viewing them as deserving of respect. However, Taylor rejects the idea that species or ecosystems as wholes might have value beyond the individual organisms that compose them. He argues that only individual living organisms are valuable.

The philosopher Holmes Rolston publishes a collection of essays called *Philosophy Gone Wild*, in which he argues that living organisms, species, systems, and natural processes all have value in themselves that human beings should recognize.

1987   The Montreal Protocol is signed by representatives of 27 nations, including the United States, committing nations to phased reductions of chlorofluorocarbon gases (CFCs) and some halon gases, which damage the stratospheric ozone layer and thereby increase solar radiation reaching the earth's surface.

Our Common Future, a report by the United Nations World Commission on Environment and Development chaired by the Norwegian Prime Minister Gro Brundtland, is published. It contains a key definition of sustainable development: "development

1987
cont.
which meets the needs of the present without compromising the ability of future generations to meet their own needs."

1988
Holmes Rolston publishes an important systematic work on environmental ethics—*Environmental Ethics: Duties to and Values in the Natural World*—in which he further develops views expressed in *Philosophy Gone Wild*.

1989
The oil tanker *Exxon Valdez* runs aground off Prince William Sound in Alaska. Substantial oil pollution results, causing increased concern about environmental damage and the effects of an oil-based society on the natural environment.

Ecology, Community and Lifestyle, a systematic book-length approach to environmental philosophy by the deep ecologist Arne Naess, is translated into English from Norwegian.

Roderick Nash publishes *The Rights of Nature: A History of Environmental Ethics*, which gives an account of the development of environmental ethics, focusing on the United States.

Founding of the International Society of Environmental Ethics (ISEE) in the United States.

1990
The International Society for Environmental Ethics publishes its first newsletter, edited by Holmes Rolston.

1991
Publication of the philosopher Lawrence Johnson's *A Morally Deep World*, in which he attempts to unite accounts of environmental ethics that focus on the individual organism and accounts that focus on ecological collectives such as ecosystems and species.

1992
The Earth Summit, an international conference, is held at Rio de Janeiro, Brazil. World leaders sign on conventions on biodiversity, climate change, forestry, and also Agenda 21, a manifesto for local and global sustainable development.

Publication of the first edition of the journal *Environmental Values*.

1993
The first survey textbook on environmental ethics, *Environmental Ethics*, by Joseph Des Jardins, is published in the United States.

1994
Establishment of the international Internet electronic academic discussion e-list *Enviroethics*.

1995 Establishment of an extensive website in environmental ethics at the University of North Texas.

1996 Founding of a new journal in environmental ethics, *Ethics and the Environment*, based at the University of Georgia. By this time, at least 26 systematic works of environmental ethics and 18 anthologies of papers in environmental ethics have been published.

CONTEMPORARY ETHICAL ISSUES

Chapter 3: Biographical Sketches

This biographical selection includes some of the individuals who have been, or who currently are, important in the development of ideas about environmental ethics. It is not intended to be exhaustive. The chapter is divided into two sections: individuals historically important in the development of environmental ethics, and key figures in the development of thinking in environmental ethics today.

Historically Important Figures in Environmental Ethics

Rachel Carson (1907–1964)

Rachel Carson is often thought of as the founder of the environmental movement in the United States. Qualified both in English (having studied English at the Pennsylvania College for Women) and in zoology (studying at the Johns Hopkins University and the University of Maryland), she became a genetic biologist in the United States Fish and Wildlife Service. In 1951, she published *The Sea around Us*, which, owing to its eloquent style, brought her national acclaim. However, she is most widely remembered for her controversial 1962 book *Silent Spring*. This book grew out of her concern about the toxic effects on wildlife and human beings of the contamination of the American countryside by synthetic pesticide, herbicide, and insecticide residues. She argued that the widespread and indiscriminate use of such chemicals, in particular chlorinated hydrocarbons and organic phosphorus insecticides, was seriously affecting ecological systems in the United States. Such chemicals, she maintained, had the potential to cause ecological catastrophe and acted as carcinogenic agents in the human environment. Furthermore, she argued, evidence suggested that many insects were evolving resistance to these chemicals, which would thus, in the long run, prove ineffective in controlling them. In place of these methods, Carson recommended the biological control of unwanted species.

The combination of scientific expertise with passionate eloquence in *Silent Spring* proved an inspiration to a fledgling environmental movement and ultimately resulted in tighter controls on pesticide use in the United States. Carson herself, however, did not live to see the long-term effect of her work. She died of cancer two years after the publication of *Silent Spring*.

Ralph Waldo Emerson (1803–1882)

Ralph Waldo Emerson was born in Boston, Massachusetts. He was the son of a Unitarian minister and, after studying at Harvard, became a Unitarian minister himself in 1829. After the death of his first wife he resigned his ministerial position and traveled to Europe, spending the year with the English Romantic writers Samuel Taylor Coleridge and William Wordsworth. Association with English and German Romantics developed Emerson's interest in religion and natural philosophy. In 1834, he moved to Concord, Massachusetts, where he became friendly with Henry David Thoreau, and together they developed what has become known as New England Transcendentalism. This movement was characterized by an emphasis on the importance of imagination over reason, an organic rather than a mechanical understanding of the natural world, and a belief in the purity of wilderness, contact with which would enable humans to develop their spiritual potential. Emerson outlined these views most prominently in his 1836 book *Nature*. They were developed in the Transcendentalist periodical *The Dial*, which was established in 1840. His collected *Essays* (2 vols., 1841 and 1844) brought him international attention, and he spent much of the rest of his life publishing poems and lectures, traveling, and lecturing.

Aldo Leopold (1887–1948)

Born in Iowa, Aldo Leopold trained at the Yale School of Forestry and began work for the U.S. Forestry Service in 1909. In 1924, he moved to

Wisconsin (where he lived for the rest of his life) to take up a position as associate director of the Forest Products Laboratory at Madison. Nine years later he became professor of wildlife management at the University of Wisconsin, Madison.

While a young forester, Leopold adhered to the traditional forestry approach to game management, attempting to wipe out game predators such as wolves. However, later reflection on this practice convinced him that such management techniques disturbed ecological systems and led to unsustainable explosions in game populations. He became increasingly concerned about the effects of many kinds of human activities on the natural environment, including inappropriate recreation and wildlife management strategies and the development of previously wild areas for industry and agriculture. He concluded that human beings needed to develop a new ethical relationship with the natural world. This relationship was explored in his major collection of essays, *A Sand County Almanac and Sketches Here and There* (1949). This work, as well as describing his own life on a farm in Wisconsin, outlines his idea of a "land ethic." His proposal, which has become famous, was that humans should see themselves as "plain members and citizens" of the ecological community; and that they should judge the merit of their actions by considering whether they promoted the "stability, integrity and beauty of the biotic community."

Leopold died fighting a fire in 1948, and *A Sand County Almanac* was published posthumously. It has been very influential in the development of more recent writing in environmental ethics, in particular that of J. Baird Callicott, Leopold's recent major interpreter.

John Muir (1838–1914)

John Muir was born in Dunbar, Scotland. His family emigrated to an 80-acre farm in Wisconsin in 1849. As a youth, Muir worked on the farm and later took up studies at the University of Wisconsin. He interrupted his studies to avoid the Civil War draft, moving to Canada and developing his skills as a naturalist and wilderness inhabitant. After the Civil War, Muir planned to walk to South America. En route in 1868, he passed through Yosemite Valley in California, which so attracted him that he remained for six years, becoming interested in the glaciology of the area. His writings at this time formed the basis for some of his later books, including his important book *My First Summer in the Sierra*. His work—in particular two articles in the magazine *Century* in 1889—was influential in the designation of Yosemite as a National Park in 1890. In 1892, Muir founded and became president of the Sierra Club. He dedicated much of his life to the protection of wilderness areas, in particular Yosemite National Park, from what he regarded as inappropriate development. In 1913, the granting of permission to improve San Francisco's water supply by flooding Hetch Hetchy, a valley within Yosemite National Park, devastated him, and he died of pneumonia a year later. He left behind him a body of writing celebrated for its understanding of ecology and wilderness living, and its affirmation that human beings are ecologically and spiritually one with the natural world.

Ernst Schumacher (1911–1977)

Born in Germany, Ernst Schumacher traveled to England to study economics at New College, Oxford, in 1930. After taking his degree, he moved to the United States, where he taught economics at Columbia University in New York. Following a spell in business and agriculture, he began his long career as an economic adviser. From 1946 to 1950, he worked as economic adviser to the British Control Commission in Germany, and from 1950 to 1970 he was the adviser for the British Coal Board. In 1973, he published *Small Is Beautiful: A Study of Economics as If People Mattered*, in which he argued that Western societies should quit pursuing environmentally and socially damaging economic strategies. He maintained that Western societies were recklessly squandering natural capital—in particular their fossil fuels—in pursuit of endless economic growth and higher standards of living. He believed that this materialistic lifestyle was not sustainable in the long term, because it was environmentally destructive, used up scarce resources, and had divisive social effects. It also ignored or sidestepped questions about the ethical implications of such economic practices for human beings and the environment. Schumacher argued instead for a "Buddhist economics," emphasizing human well-being over increased technology and the importance of smaller-scale technology and communal ownership. To this end he established and subsequently became chairman of the Intermediate Technology Development Group, which helped to establish local, small-scale and human-centered enterprises in developing countries. He was awarded the Companion of the British Empire for his work in 1974.

Albert Schweitzer (1875–1965)

Born in Alsace, Germany (now in France), Albert Schweitzer became well known as a musician and theologian, publishing a book on Bach in 1905 and an important theological work, *The Quest of the Historical Jesus*, while a lecturer at Strassburg University in 1910. He gave up his academic career to train as a medical missionary, qualifying as a medical doctor in 1913. He then moved to Lambarene in the Gabon province of French Equatorial Africa, where he founded a hospital for local people on the Ogowe River. He returned to France during World War I and was interned as a prisoner of war. In 1924, he returned to Lambarene, where he increased the capacity of the hospital by incorporating a leper colony. He was awarded the Nobel Prize for Peace in 1952.

During his time at Lambarene, Schweitzer wrote extensively about social and philosophical questions. He is most well known for his work *The Philosophy of Civilisation* (2 vols., 1923). In this work, Schweitzer argued for a "new way" in European philosophy, a way based on the recognition of the principle of "reverence for life"—the expression for which his work is best known. He argued that all living organisms—humans, animals, plants—have a "will-to-live," a desire to move toward self-realization and toward unification with other living beings. As humans recognize the need to fulfill their own will-to-live, so they should show reverence for the will-to-live of all other living things, regarding all of their lives as sacred. For this reason, Schweitzer urged that humans should forgo all unnecessary killing and seek to make amends for any harms they may have caused.

Although Schweitzer's philosophy was not widely followed in his lifetime, his understanding of reverence for life has since been significant in influencing the development of work in environmental ethics. In particular, his insistence on reverence toward all living organisms has been taken up within some sections of the deep ecology movement.

Henry David Thoreau (1817–1862)

Henry David Thoreau was born in Concord, Massachusetts, and studied at Concord Academy and Harvard. After graduating in 1837, he ran a school in Concord with his brother until he decided to become a writer in 1840. He supported himself by making lead pencils and gardening and became the gardener in Ralph Waldo Emerson's household in 1841. He became associated with Emerson's Transcendentalist movement and helped Emerson edit his periodical, *The Dial*. Emerson encouraged Thoreau to keep journals, and these journals formed the basis of Thoreau's published writing. Best known of these writings is *Walden* (1854), an autobiographical essay written some years earlier while Thoreau was living in isolation and self-sufficiency in a hut he had built at Walden Pond near Concord. In *Walden*, Thoreau emphasized the importance of living a self-directed life of simplicity and harmony with the natural world. He understood humans to be part of, rather than apart from, nature, and argued that spending time in wilderness areas provided humans with insight into their own spiritual reality.

Thoreau remained at Walden Pond for just over two years before returning to live in town. In 1849, he published his influential work *On the Duty of Civil Disobedience* after being imprisoned for not paying poll tax to a government he perceived as supporting war and slavery. His reflection on the duty to disobey unjust laws influenced Mahatma Gandhi and Martin Luther King Jr. and has subsequently been an important document for environmental campaigners throughout the world who have practiced civil disobedience. Despite the later popularity of both *Walden* and *On the Duty of Civil Disobedience*, Thoreau's work was largely unrecognized during his lifetime. He died of tuberculosis when he was only 45.

Alfred North Whitehead (1861–1947)

The son of an Anglican vicar, Alfred North Whitehead was born in Kent, England. In 1880, he began a degree in mathematics at Trinity College, Cambridge, where he remained as a lecturer until 1910. Between 1910 and 1913, together with his former pupil Bertrand Russell, he published *Principia Mathematica*, a study of the relationship between mathematics and philosophical logic.

Whitehead became increasingly interested in philosophy, and in 1920 published a book on his philosophical understanding of nature, *The Concept of Nature*. This interest in philosophy led him to accept a post in philosophy at Harvard in 1924. He then published a series of books exploring his understanding of nature, most prominently *Science and the Modern World* (1925) and *Process and Reality* (1929). In these works he developed what he called his "philosophy of organism." He argued that at the most fundamental level the universe was composed from "actual entities"—minute, constantly changing pulses of feeling. Following from this, Whitehead argued that all the elements of the universe were organically interrelated and that no firm distinctions could be drawn between living and nonliving matter. He also argued that value could be found everywhere and was not confined to human beings.

Since Whitehead's philosophy was expressed in difficult language, interest in it has been confined, until recently, to a small group of philosophical scholars. However, some environmental philosophers and theologians, most notably Charles Hartshorne and John Cobb, have now turned back to his work, believing that his understanding of nature and value can provide a fruitful way of approaching environmental philosophy.

Current Contributors to Work in Environmental Ethics

Murray Bookchin (b. 1921)

Murray Bookchin was born in New York. His interest in environmental problems was first expressed in 1951, when he published an article about the effects of man-made chemicals on human beings and the environment. The radical nature of his thought was developed during the 1960s, when he began to publish critiques of urban policy in the United States. However, it is his work during the 1980s in establishing a school of thought called "social ecology" for which he is best known. In 1980, Bookchin published *Towards an Ecological Society*. In this and in his subsequent book, *The Ecology of Freedom*, he argued that "oppression" and destruction of the natural world should be seen as a direct result of the oppression within human societies. This oppression, he argued, stems from the hierarchical construction of human societies, in which some humans have the power to dominate and control others. Societies that control and dominate people, he argued, will inevitably control and dominate nature. He maintained that the resolution of environmental problems—which he regarded as a central issue—was therefore dependent on structural changes in human society.

Bookchin's analysis, coming from a left-wing, anarchistic tradition, has made an important contribution to environmental thought and become the starting point for a variety of approaches to environmental ethics. However, his stress on the significance of *human* society, his rejection of policies of population control, and his strongly expressed dislike for what he called the "eco-la-la" associated with some more esoteric parts of the environmental movement has more recently led him into disputes with deep ecologists and environmental ethicists who are rooted in deep ecological traditions.

Andrew Brennan (b. 1945)

Andrew Brennan was born in Kirkcaldy, Scotland. He graduated in logic and metaphysics and moral philosophy from the University of St. Andrews in 1967 and went on to take higher degrees in philosophy from the University of Calgary and the University of Oxford. After teaching for most of his career at the University of Stirling, U.K., Brennan moved to take up the chair in philosophy at the University of Western Australia in January 1992.

His first environmental ethics publication was "The Moral Standing of Natural Objects" (*Environmental Ethics* 1984), which argues that their lack of an intrinsic function gives natural objects and systems a special claim for moral consideration. His book *Thinking about Nature* (1988) links a humanist environmental ethic with an analysis of scientific ecology. It advocates ethical pluralism and gives attention to practical policy issues. In later publications he has continued to defend moral pluralism and has been critical of the centrality given to wilderness preservation by other authors.

He was a founding board member of the International Society for Environmental Ethics and is on the editorial boards of several journals, including *Environmental Ethics*, *Environmental Values*, and the *Journal of Applied Philosophy*. He edits the Routledge series *Environmental Philosophies* and was also the editor of a volume for the International Research Library of Philosophy, *The Ethics of the Environment* (1995).

J. Baird Callicott (b. 1941)

Born in Memphis, Tennessee, J. Baird Callicott graduated from Messick High School in 1959 and majored in philosophy at Rhodes College. He graduated with honors in 1963 and moved to Syracuse University, where he took an M.A. in 1966 and a Ph.D. in 1971. He began to teach at the University of Wisconsin, Steven's Point, where in 1971 he offered the world's first undergraduate course called Environmental Ethics. In 1979, his paper "Elements of an Environmental Ethic: Moral Considerability and the

Biotic Community" was published in the first edition of the journal *Environmental Ethics*.

In 1980, in volume 2 of *Environmental Ethics*, Callicott published a paper called "Animal Liberation: A Triangular Affair." Influenced by the work of the writer and forester Aldo Leopold, Callicott attacked the common assumption that environmental ethicists and animal liberationists share similar beliefs. He argued that whereas animal liberationists are primarily concerned with the suffering and well-being of individual animals, environmental ethicists are interested in the whole biotic community or system. At the system level, the well-being of individuals is far less significant and pain is of little concern. This important paper divided animal liberationists from environmental ethicists, a division that has, to some extent at least, persisted over time. Callicott has himself tried to reconcile the two groups by publishing a paper in the journal *Between the Species* (1988) entitled "Animal Liberation and Environmental Ethics: Back Together Again."

In addition to these key papers, Callicott has played a major role in the development of environmental ethics in the United States in the 1980s and 1990s. He has published a substantial amount on Leopold, editing and contributing to an essay collection called *Companion to a Sand County Almanac* (1987); he has also written a large number of articles on environmental ethics. In 1994, he published a book-length study, *Earth's Insights: A Survey of Environmental Ethics from the Mediterranean Basin to the Australian Outback*. Recently he has been working on the nature and significance of wilderness areas in the United States. He moved to the Center for Environmental Philosophy at the University of North Texas in 1995.

Eugene Hargrove (b. 1944)

Eugene C. Hargrove was born in Detroit, Michigan, but grew up in St. Louis, Missouri. He received his M.A. (1968) and Ph.D. (1974) in philosophy from the University of Missouri. After a year of postdoctoral work on Wittgenstein, Hargrove became a Rockefeller Foundation fellow in environmental affairs. Building on his experiences as an environmental activist for the National Speleological Society (1971–1974), he pursued original research into the history of ideas behind environmental thought as it pertains to environmental ethics or environmental philosophy. Hargrove founded the journal *Environmental Ethics* in 1978 at the University of New Mexico, the first issue of which appeared in January 1979. It was the first, and still is the most important, journal in the field. In 1980, he created a nonprofit organization, Environmental Philosophy, Inc., to own and manage the journal and moved it to the University of Georgia in 1981. In 1989, he created the Center for Environmental Philosophy and moved it and the journal to the University of North Texas, where he developed a graduate program focused entirely on environmental philosophy and created Environmental Ethics Books, a reprint books series for important books in the field.

Hargrove is the author of *Foundations of Environmental Ethics* (1989), based in part on his historical research as a Rockefeller Foundation fellow, and the editor of three anthologies: *Beyond Spaceship Earth: Environmental Ethics and the Solar System* (1986), original papers dealing with ethical and environmental issues related to the U.S. space program; *Religion and Environmental Crisis* (1986), original papers on religion and the environment; and *The Animal Rights/Environmental Ethics Debate: The Environmental Perspective* (1992), previously published articles and chapters from books discussing the relationship of environmental ethics to animal liberation and animal rights. In 1995, Hargrove developed a comprehensive system of electronic web pages on environmental ethics that link to centers of importance in environmental ethics internationally (accessible at http://www.cep.unt.edu). Currently, he is concerned with finding ways to introduce environmental ethics into schools and into the graduate programs that train environmental professionals. His role as facilitator for work in environmental ethics, both in the United States and internationally, continues to be extremely significant.

James Lovelock (b. 1919)

James Lovelock was trained in biology and medicine at Manchester and London, U.K. From 1941 to 1961, he worked at the National Institute of Medical Research. In 1961, he was invited to work with the National Aeronautics and Space Administration (NASA) to develop instruments to explore the surface and atmosphere of the moon and other planets. After leaving NASA, Lovelock chose to work largely independently as a research scientist, setting up his own laboratory in Cornwall, England, where he still lives. He has pursued a distinguished scientific career, including the publication of early work on the persistence of chlorofluorocarbons in the earth's atmosphere and the invention of the electron capture detector.

Lovelock is, however, chiefly renowned in the environmental movement for the development of what he labeled "the Gaia hypothesis," first published in his book *Gaia* in 1979. In this book he argued that the earth in its entirety, including rocks and atmosphere, can be regarded as a single living organism, and that the living organisms on earth can control their nonliving environment to make it comfortable for life. Lovelock developed this hypothesis, and responded to criticisms of it, in his 1986 book, *The Ages of Gaia*, where he considered in more detail how "Gaia" may have evolved and what the philosophical and ethical implications of this hypothesis might be.

Although the Gaia hypothesis is still controversial among scientists, Lovelock's work is widely respected, especially in Britain, where he has been made a fellow of the Royal Society.

Mary Midgley (b. 1916)

Mary Midgley was educated at Downe House, near Newbury, England. She studied at Somerville College, Oxford, and was awarded a first-class honors degree. During World War II she worked briefly for the Ministry of Production, then spent time as a classics teacher. After the war she researched in Oxford before taking a position as a philosophy lecturer at Reading University. Following a career break to bring up a family, she worked in the philosophy department at the University of Newcastle, U.K. She began to publish in the field of human nature and human understanding of and relationships with other animals. Her most prominent books are *Beast and Man: The Roots of Human Nature* (1978) and *Animals and Why They Matter* (1983). She has also published extensively on the place of scientific thinking in modern society, in particular in her works *Evolution as a Religion* (1985) and *Science as Salvation* (1992).

In both these (interconnected) areas, Midgley's work has been of significance. She has become well known for bringing together insights from science and philosophy to address questions concerning human beings, animals, and the environment in a characteristically clear and forthright style. In addition, she has made a number of contributions to the development of ideas in animal and environmental ethics, most prominently her suggestion that human beings and domesticated animals live in a "mixed community," a situation that may create moral obligations for humans—a suggestion developed by J. Baird Callicott, among others.

Although Midgley has now retired from her position at Newcastle University, she is still actively working and publishing in the field of environmental philosophy.

Arne Naess (b. 1912)

Born in Oslo, Norway, Arne Naess studied as an undergraduate at the Sorbonne in Paris, then undertook graduate work in Vienna, Berkeley, and Oslo. He received his doctorate from Oslo University in 1938 and was appointed professor of philosophy there in 1939. His subsequent distinguished philosophical career includes publications on a range of topics from philosophy of science and philosophy of language to the nature of democracy and the philosopher Spinoza. He also served as the editor of the international philosophical journal *Inquiry*, which he founded in 1958.

In addition to these philosophical interests, Naess was fascinated by wilderness and was an accomplished mountaineer. During the 1960s, he became active in the growing environmental movement. This interest led to the publication in 1973 of his article "The Shallow and the Deep, Long-Range Ecology Movement" in *Inquiry*, a publication that is widely regarded as marking the beginning of the modern deep ecology movement. In this article, Naess distinguished between shallow "environmentalists," who are primarily concerned with pollution and resource depletion, and the more profound concerns of deep ecologists—issues such as egalitarianism, diversity, intrinsic value in nature, and decentralization. Naess's subsequent writing has developed these themes, put forward in most detailed form in his book *Ecology, Community and Lifestyle* (1989). In this book, he describes his own personal "ecosophy"—the set of assumptions and values that underpin his interactions with the natural world—and encourages others to develop their own ecosophy.

Now retired, Naess still travels and publishes widely. His work has influenced many in environmental ethics, including those ethicists who ultimately reject a deep ecology approach.

Tom Regan (b. 1938)

Born in Pittsburgh, Pennsylvania, Tom Regan studied at Thiel College and the University of Virginia, where he obtained his doctorate in philosophy. During the 1970s, he began to publish work on animal rights, an interest that grew during the 1980s with the publication of a series of books, articles, and edited essay collections on animal rights and also on environmental ethics.

Among the most important of these works is his 1982 essay "The Nature and Possibility of an Environmental Ethic," published in a collection he edited: *All That Dwell Therein: Animal Rights and Environmental Ethics.* In this essay, Regan explores the possibility that there might be a kind of "inherent goodness" in natural objects that should generate "admiring respect" in humans and lead them to adopt a "preservation principle" of "nondestruction and noninterference."

In his subsequent work, Regan has focused less on environmental ethics and more on animal rights. His book *The Case for Animal Rights* (1983) is widely regarded as the classic book in the field, alongside Peter Singer's *Animal Liberation.* In this book, Regan argues that, if it is agreed that human beings have rights, there are no logical grounds on which to exclude animals (or to be more precise, adult mammals). He maintains that therefore animals also have rights, and humans correspondingly have duties to respect them. In adopting this rights approach to animal ethics, he rejects the work both of Enlightenment philosophers like Descartes (who maintained that animals were little different from machines) and utilitarian philosophers like Peter Singer, who argue that animals are morally significant because they can feel pleasure and pain.

Regan is currently a professor at the University of North Carolina. His work in animal rights is recognized as important in influencing a whole generation of academic and pressure-group activity. Since the publication of *The Case for Animal Rights*, he has written extensively about changing attitudes toward animals and the environment, in particular in his 1991 book *The Thee Generation: Reflections on the Coming Revolution.*

Holmes Rolston III (b. 1932)

Holmes Rolston III was born in the Shenandoah Valley, Virginia. His richly varied educational career included studying physics as an undergraduate at Davidson College, then entering theological seminary and completing a Ph.D. in theology at Edinburgh University, Scotland, in 1958. He then worked for some years as a Presbyterian pastor before taking a master's degree in philosophy of science at the University of Pittsburgh. An academic appointment in philosophy followed at Colorado State University, where he became a full professor in 1976.

Rolston has been of central importance to the development of environmental ethics as an academic discipline, both as a profuse writer in the field and as one of the founders of the journal *Environmental Ethics*. He has published widely in environmental ethics, including three important books: *Philosophy Gone Wild* (1986), *Environmental Ethics* (1988), and *Conserving Natural Value* (1994). Rolston argues that the natural world carries intrinsic values that human beings should recognize. These values exist not only at the level of individual organisms but also in species, ecosystems, and natural processes. The existence of such values means that humans have duties toward the natural world, including duties to protect species and ecosystems from destruction. Besides publishing in environmental ethics, Rolston has also written in philosophy of science and religion more generally, including his 1987 book *Science and Religion: A Critical Survey*.

Rolston is associate editor of the journal *Environmental Ethics* and serves on the editorial boards of a number of other journals, including *Environmental Values*. He currently holds the position of University Distinguished Professor of Philosophy, Colorado State University.

Kirkpatrick Sale (b. 1937)

Born in Ithaca, New York, Kirkpatrick Sale studied history at Cornell University, graduating in 1958. He began work as a journalist, initially for the left-wing journal *New Leader*. After spending some time in Africa, he returned to the United States, where he left journalism and began to work as a freelance writer, publishing work on American radical politics. This work radicalized Sale, and in 1980 he published *Human Scale*, a critique of the nature of modern technological, industrial societies. In this book he argued that such societies are facing a profound crisis that requires change not in the kinds of technology but in the fundamental value systems by which they operate. He rejected what he saw as the dominant idea of Western industrial culture—"bigger is better"—and argued instead that human societies should operate on a smaller, human scale.

This critique led Sale to engage increasingly with environmental ideas, in particular with concerns about the limited resources and lack of sustainability of industrial societies. In this context, Sale developed the idea of bioregionalism, expounded in his 1985 book *Dwellers in the Land: The Bioregional Vision*. Bioregionalism is based on the ethical principle that human individuals and communities should live in harmony with the natural world, adapting to the existing physical and ecosystemic patterns in a particular region rather than trying to change them. Sale's bioregional vision has proved an inspiration to many in the environmental movement, encapsulating the kind of society that might result from the application of a variety of approaches to environmental ethics—including some forms of deep ecology. More recently, Sale has published *Conquest of Paradise: Christopher Columbus and the Columbian Legacy* (1990). This book explores the effects of the introduction of European thinking, in particular European ethical attitudes toward the environment, into a continent inhabited by Native Americans. With this work, as well as his writing on bioregionalism, Sale has contributed to the development of environmental ethics in its widest sense.

Peter Singer (b. 1946)

Peter Singer was born in Melbourne, Australia. As an undergraduate, he studied at the Scotch College, University of Melbourne, before moving to England to work on a doctorate at University College, Oxford. He was Radcliffe Lecturer at Oxford from 1971 to 1973 before working at New York University and La Trobe University. In 1977, he was appointed professor of philosophy at Monash University, where he has been based since.

Singer's main philosophical work has always been in political and moral philosophy. While at Oxford in the early 1970s, he became interested in human ethical obligations toward animals and the question of vegetarianism. This interest resulted in the publication of a work called *Animal Liberation*, first as an article in the *New York Review of Books* in 1973, and later in 1975 as a book. *Animal Liberation* was written from within a *utilitarian* tradition in philosophy, where the fundamental moral principle might be summarized as the maximization of happiness and the minimization of pain and suffering. Singer argues that because animals are, like humans, able to suffer, their interests in avoiding suffering should be taken into account in moral decision making. Indeed, he argues that equal moral consideration should be given to equal interests, whatever the species of being involved. Failure to take other species into account is *speciesism*, he maintains, a concept similar to racism and sexism. Much of *Animal Liberation* is taken up by descriptions of speciesism in practice—in factory farming and animal experimentation. *Animal Liberation* has been hugely significant in the subsequent popular explosion of interest in human ethical relationships with animals. However, it has also been important in environmental ethics as a key to opening up debate about the moral significance of nonhumans.

Singer developed and refined his views on animals in later writing—especially in his book *Practical Ethics* (1979). He has since published widely on a range of ethical issues, including abortion, euthanasia, and biotechnology.

He is currently director of the Center for Human Bioethics at Monash University, Australia.

Paul Taylor (b. 1923)

Paul Warren Taylor was born in Philadelphia, Pennsylvania. He received a B.A. in 1947, an M.A. in 1949, and a Ph.D. in 1950, all from Princeton University. He then moved into an academic position at the City University of New York, where he was appointed to a professorial chair in philosophy in 1967. During his time at City University of New York, Taylor published a number of books on ethics, most prominently *Normative Discourse* (1961) and *Principles of Ethics: An Introduction* (1975). He also edited several collections of essays on ethics.

In 1981, Taylor published an essay entitled "The Ethics of Respect for Nature" in the third edition of the journal *Environmental Ethics*. He developed and expanded these arguments in his important contribution to the field of environmental ethics, *Respect for Nature: A Theory of Environmental Ethics* (1986). In this work, Taylor argues that humans should adopt an attitude of "respect for nature," which involves the acknowledgment that humans are part of an interconnected and interdependent ecosystem composed from individual organisms all pursuing their own good in their own way. Because all individual organisms do pursue their own good in their own way, Taylor argues that they have *inherent value*. Those who have adopted an attitude of respect for nature should respect this inherent value, avoiding harmful behavior toward all living organisms. One of the striking parts of Taylor's argument is that respect is owed to all individual living organisms equally (whether animal or vegetable). Recognizing the difficulties of such a position, Taylor devotes much of his book to the development of a series of priority principles to help enact this theory in practice.

Though regarded critically by environmental ethicists who work at the level of the ecosystem or species rather than the individual organism, *Respect for Nature* has been one of the key systematic works in environmental ethics. For this reason Paul Taylor, who retired in 1990, should be regarded as one of the central figures in the development of environmental ethics.

CONTEMPORARY ETHICAL ISSUES

Chapter 4:
Major Issues in Environmental Ethics

In a field as large and as complex as environmental ethics, an alphabetical listing such as this has its shortcomings. It must be selective, inevitably omitting some areas of concern. In addition, topics overlap and material must be divided, almost arbitrarily, between two entries. Where relevant material can also be found under other entries, cross-references direct the reader to it.

For each entry, a definition and a general overview of the topic is presented (and in some cases further information, such as international policy relating to it) before its relevance to environmental ethics is considered. Each entry concludes with a list of the references used to write the entry, which the reader may consult for more detailed information.

Agriculture

Agriculture can be defined as the art or practice of cultivating of the land, including the domestication of plants and animals, usually for purposes of production for human use.

Agriculture is thought to have been first practiced about 10,000 years ago, succeeding hunter-fisher-gatherer lifestyles in many parts of the world. It has become the dominant source of food and is also important for the production of many fibers and materials. World agricultural output and food production have steadily increased since the end of World War II in 1945, although this increase has generally been faster in more developed than in less developed countries. There has also been a rapid international expansion of aquaculture (the managed cultivation of fish and seafood), both in fresh water and along shorelines. In 1990, 13 million tons of fish and seafood were produced by aquaculture, a figure projected to rise to 25 million tons by 2010 (UNEP 1993).

However, these production increases have come at an environmental price. More intensive farming of land has led to the increased use of pesticides, herbicides, and artificial fertilizers, causing problems with residues in food and water, soil degradation, and harm to wildlife. It has also led in some areas to overgrazing and increased soil erosion. The extension of agriculture into previously wild areas has led to the loss of biodiversity and damage to ecosystems (in particular in areas of tropical forests). Aquaculture has led to the pollution of inshore and fresh waters. The pressure to increase yields and grow crops in areas previously considered unsuitable has resulted in greater water consumption for irrigation, leading to lower water tables, the construction of ecologically harmful dams, the diversion of rivers, and the loss of wetlands. Similar pressures have led to an increase in the use of biotechnology in agriculture, which has raised a variety of ethical concerns (*see* Genetic Engineering). All of these issues have led, in some developed countries, to heightened interest in organic agriculture (agriculture that uses no artificial chemicals at all) and in sustainable agriculture. Sustainable agriculture (the term most commonly used in agricultural policy) is variously defined—sometimes simply as "reduced chemical farming"; sometimes more generally as an approach to farming that is labor-intensive, reduces chemicals, practices crop rotation, and avoids soil erosion (Rawson 1995).

Sustainable Agriculture in the United States In 1962, the publication of Rachel Carson's book *Silent Spring* aroused considerable concern about the use of herbicides and pesticides on U.S. farms. In fact, since the 1950s, private research institutes in the United States had been working on low- and no-chemical input farming. In the 1970s, Iowa, California, and Minnesota began to support research programs on sustainable agriculture. In 1972, the U.S. Department of Agriculture (USDA) launched the Integrated Pest Management scheme, aimed at helping farmers to develop nonchemical ways of controlling pests (a scheme that received $200 million in funding in 1995). This program was followed by a series of amendments to the Clean Water Act in 1977, which promoted programs of soil conservation to bring water quality benefits. In 1988, the federal government established the Low Input Sustainable Agriculture Program (LISA), renamed the Sustainable Agricultural Research and Education Program (SARE) in the Food, Agriculture,

Conservation and Trade Act of 1990. This program provides grants for experiments in reduced-chemical farming and funds the dissemination of information through organizations like the Sustainable Agriculture Network. Since 1990, there have been a range of federal initiatives to promote and fund projects reducing pesticide and fertilizer use involving the USDA and the Environmental Protection Agency.

Relevant Ethical Issues   Agriculture has been a topic little discussed by academic environmental ethicists, and it is not difficult to understand why. Environmental ethicists have focused debate on whether the environment has value irrespective of its usefulness to human beings; issues arising from this controversy have been the preservation of wilderness and protection of biodiversity. Agriculture is, by definition, about the *use* of land for human benefit, and such land can no longer, of course, be considered to be wilderness. Thus, agriculture has tended to fall outside the discussions of environmental ethicists; as an exploitative process that destroys wild areas and reduces biodiversity, it has even been the focus of some hostility.

Even so, agriculture clearly raises a number of issues relevant to environmental ethics. These have received some discussion, most comprehensively in Paul Thompson's 1995 book, *The Spirit of the Soil: Agriculture and Environmental Ethics*. Thompson points out that most ethical discussion about agriculture and the environment has worked with the (utilitarian) assumption that the purpose of agriculture is to provide the greatest benefit to the largest number of people. The benefit is usually seen in terms of the greatest possible production with fewest human costs. From within this framework, the ethical acceptability of pesticides, herbicides, fertilizers, the release of genetically modified organisms, and so on can be assessed. If their benefits (in terms of increased yield) outweigh their costs (in terms of harm to human health, destruction of life-supporting ecosystems, etc.), then such technologies should be pursued. If their costs outweigh their benefits, then the technology should be either rejected or reshaped so that the costs are minimized and the benefits maximized. This kind of debate, Thompson rightly argues, has dominated the recent ethical discussion about agricultural practices and the environment.

However, as Thompson and other writers point out, the costs-versus-benefits approach is not the only way environmental ethicists can think about the rights and wrongs of agriculture. Another long-standing approach, frequently traced back to Genesis in the Bible, is the idea that farmers are stewards or custodians of the land. This stewardship role (often understood in a theological context) undercuts the idea that farmers (or indeed anyone else) *own* the land. Rather, they take care of it on behalf of God and are responsible to God for its good management. (Nontheological versions of this approach usually suggest that farmers take care of the land on behalf of future generations of humans.) This understanding of agriculture rejects a short-term, profit-focused, and potentially abusive relationship with the land, preferring to advocate values of nurture rather than exploitation (a view often associated with the American writer Wendell Berry).

Advocacy of a stewardship environmental ethic is often closely associated with the rejection of "giantism" and the power of corporate agribusiness, emphasizing the importance of the family farm and the stability of traditional close-knit agricultural communities.

Some environmental ethicists have rejected both the utilitarian and the stewardship understandings of agriculture. They argue that both are anthropocentric in their approach because they are concerned purely with the benefits to human beings. The "common good" of the utilitarian approach, they point out, aims at an exclusively *human* common good. The stewardship approach, viewed from this perspective, is fundamentally hierarchical: God owns the land, humans farm it, and the land is solely viewed as a place for human production. There is no sense that the land, or elements of it, may have value beyond human use. (For detailed critiques of stewardship, see Ebenreck 1983, Palmer 1992, and Thompson 1995.) For such ethicists, affirming the value of the land beyond its importance for human productive purposes is very important. These affirmations need not mean that the practice of agriculture is rejected altogether, but rather that it is carried out with respect for nonproductive ecological values and that some places are protected altogether from agriculture.

A variety of interpretations of such ecological values exists. Most commonly, philosophers have based their work on the land ethic of the forester Aldo Leopold, who said that "a thing is right when it tends to preserve the stability, integrity, and beauty of the biotic community. It is wrong as it tends otherwise" (Leopold 1949). Although an extreme interpretation of this principle might suggest that all agriculture should be rejected, Leopold himself did not take such a view. Those working in Leopold's tradition, such as the philosopher J. Baird Callicott, have argued that agriculture should be *holistic*, taking into account its effect on the entire biotic community as well as on the human community. Such an approach moves beyond either utilitarian or stewardship environmental ethics.

Currently, there is little agreement among environmental ethicists about how agriculture should be regarded. It is vital to human survival (and increasingly to that of many animals); at the same time, it disturbs and damages wild ecosystems, by necessity. The attempt to reconcile these two elements will likely dominate writing about agriculture and environmental ethics in the future.

References: Berry, Wendell. 1977. *The Unsettling of America: Culture and Agriculture.* New York: Harcourt Brace Jovanovich.

Callicott, J. Baird. 1990. "The Metaphysical Transition in Farming: From the Newtonian-Mechanical to the Eltonian Ecological." *Journal of Agricultural Ethics* 3(1): 36–49.

Ebenreck, Sara. 1983. "A Partnership Farmland Ethic." *Environmental Ethics* 5(1): 33–47.

Leopold, Aldo. 1949. *A Sand County Almanac.* Oxford: Oxford University Press.

Palmer, Clare. 1992. "Stewardship: A Case Study in Environmental Ethics." In Ian Ball et al., *The Earth Beneath.* London: SPCK.

Rawson, Jean. 1995. "Sustainable Agriculture." Congressional Research Service Report for Congress. 15 October, 95-1062.

Thompson, Paul. 1995. *The Spirit of the Soil: Agriculture and Environmental Ethics*. London: Routledge.

United Nations Environmental Programme. 1993. *The World Environment 1972–1992*. London: UNEP/Chapman and Hall.

Atmospheric Pollution

The atmosphere is the envelope of gases that surrounds the earth. Atmospheric pollution occurs when the atmosphere is contaminated by substances (whether gases or particulates) that are, from a human perspective, undesirable.

A wide variety of substances contributes to atmospheric pollution. Some of them are not of human origin, such as gases emitted by volcanic eruptions. However, most air pollution is caused by human beings and gives rise to a variety of different problems. Chief among them are stratospheric ozone depletion; acid rain; photochemical smog and poor ground-level air quality, especially in urban areas; and indoor air pollution. (For a discussion of the problem of enhanced global warming, see Climate Change.)

Ozone (O_3) is a gas that occurs naturally in the earth's stratosphere, about 25 kilometers above the surface of the earth. It filters ultraviolet (UV) light from the sun and protects life on earth from its potentially hazardous effects. That ozone depletion might be occurring was first suggested by scientists in 1974. When a hole was discovered in the ozone layer over the Antarctic in 1984, ozone depletion became a topic of international urgency. Stratospheric ozone depletion has subsequently been observed globally except over the tropics, although it is most concentrated in the cold air of the Arctic and Antarctic. The cause of this depletion was found to be a group of chemicals called halocarbons, particularly chlorofluorocarbons (CFCs). CFCs were used globally as aerosol propellants, foam-blowing agents, and refrigerants, and were previously thought to be inert and nontoxic. Effects of stratospheric ozone depletion, and hence increased exposure to UV light, include an increase in human skin cancer and cataracts, suppression of the human immune system, reduced plant growth, damage to aquatic organisms, and degradation of some materials, including some plastics.

Acid rain was first noted by the Swedish biologist Svante Oden in the 1960s. He maintained that the decline in fish populations in Swedish lakes and rivers was due to acidification of the water by acid precipitation (Oden 1968). Further investigation confirmed his findings. Acid precipitation (usually rain but also hail, snow, and fog) is caused by chemical reactions in the atmosphere when gases such as sulfur dioxide, nitrogen oxides, and ammonia mix with water to form dilute sulfuric and nitric acid. Sulfur dioxide, the most serious problem, is primarily released by coal-burning power stations, especially in those areas (such as the United Kingdom) where coal is high in sulfur. The resulting acid precipitation may well fall hundreds of miles from the point of release, as in the case of Sweden, which is largely polluted by acid rain from the United Kingdom. The environmental effects of acid rain can be devastating: it can kill all life in rivers and lakes; cause serious damage and even death to forests (in Germany the term for acid rain, *Waldsternben*, means "forest dieback"), and erode buildings. Methods do exist to remove or reduce the amount of sulfur and nitrogen oxides released by the burning of fossil fuels in power stations, but the technology is expensive and some governments are reluctant to install it.

Urban air quality has been problematic for many years, primarily due to emissions of sulfur dioxides, nitrogen dioxide, carbon monoxide, particulates from the burning of fossil fuels, and the formation of ground-level ozone. In 1990, Organization for Economic Cooperation and Development figures estimated that humans released 99 million tons of sulfur oxides, 57 million tons of particulates, 177 million tons of carbon monoxide, and 68 million tons of nitrogen oxide into the atmosphere worldwide. In recent years, the automobile has become the single major source of such emissions in urban areas. At high concentrations, these emissions can cause a variety of respiratory problems, especially in certain weather conditions. (In 1994, Los Angeles had 159 so-called unhealthy air days.) One of the major problems they create (which may not be confined to urban areas) is tropospheric or low-level ozone, a main constituent of photochemical smog. This smog can cause damage to both human health and the health of other animals, to vegetation, and to buildings. It has been implicated in the declining health of some European forests. Various measures have been introduced in an attempt to improve urban air quality, including controls on the pollutant content of fuels, emission controls, traffic management schemes, public information campaigns, and pollution alert systems.

Indoor air pollution has a variety of causes and takes a variety of forms. It includes exposure to substances such as asbestos, the entrapment of radon gas in modern airtight buildings, the emission of formaldehyde from insulating foam, nitrogen dioxide from gas and kerosene heaters, cigarette smoke, and polyaromatic carbons from burning (such as from woodstoves). These substances cause a variety of human health disorders but are not usually damaging to the environment.

Atmospheric Pollution and International Law   The nature of atmospheric pollution means that it is often transboundary (meaning that it can pass from one country to affect another) or global. Thus, although most countries, including the United States, have a range of laws controlling emissions to the air, some of the most important legal agreements are international. The first important international agreement was the Convention on Long-Range Transboundary Air Pollution, signed by 34 European and North American nations in 1979. A number of related agreements have followed. Most significant among these are the Thirty Percent Protocol, signed by 21 nations in 1985, and the NOx Protocol, signed by 27 countries in 1988. The first entailed a commitment to reducing sulfur emissions by at least 30 percent of 1980 levels by 1993; the second a commitment to reducing 1994 nitrogen oxide emissions to 1987 levels (UNEP 1993).

However, the most important international agreements on air pollution have concerned the protection of the stratospheric ozone layer. This issue was first addressed in 1985 by the Vienna Convention for the Protection of the Ozone Layer (ratified by 19 nations) and later by the 1987 Montreal Protocol on Substances That Deplete the Ozone Layer. Signatory governments were required to regulate production and consumption of CFCs and halon gases; in later amendments, signatory nations agreed to phase out production of both kinds of chemicals by 2000. Currently, more than 100 nations have ratified the Montreal Protocol.

Relevant Ethical Issues   Clearly, atmospheric pollution of the kind described above raises a number of difficult ethical issues. Atmospheric pollution can cause a variety of harms to human beings and human health, to other animals, to vegetation and marine life, to ecosystems, and, potentially, to species. But the processes that cause air pollution—driving automobiles, producing energy, keeping food cold—provide benefits of varying degrees of importance to those who use them, as well as providing employment and economic prosperity. Thus, the question of costs versus benefits is a central ethical issue. Other questions are also raised: Who is ultimately responsible for producing this pollution, and who should pay for it (both financially and in terms of lost amenities)? In the case of polluting, coal-fired power stations, for instance, who should be responsible for paying for pollution-abatement equipment? The power generation companies, even though spending on this may mean redundancies and a less efficient energy production service? Consumers, who ultimately demand the energy, even though for some, such as the elderly, this burden may mean an increased risk of death from hypothermia? The government, even where this means higher taxation, or the switching of funding from other important areas of spending, such as health or education? And what if the pollution does not affect the country of origin, as when Britain exports acid rain to Sweden, or when tropical countries largely unaffected by ozone depletion produce CFCs? What international ethical responsibilities do nations have? All of these issues are ethically problematic.

Clearly, the harms caused by atmospheric pollution can be understood in different ways. For instance, a philosopher who argues that humans—and animals—have rights might maintain that ozone depletion and poor urban air quality violate the rights of these individuals to live a healthy life and is therefore wrong. However, this argument runs into difficulties in many cases because of the diffuse nature of atmospheric pollution. Having a right to something (in this instance, to a healthy life) usually means that other individuals have a duty to protect that right. But in this case, exactly who has the duty? Is it the individual polluter? The individual whose right is being violated may be a polluter him- or herself. And it is reasonable to argue that no one individual's pollution is causing the violation of rights but rather the combined pollution created by many individuals. The right to a healthy life seems to be violated by a group, then, rather than an individual. So who has the duty to protect the right? This difficult question has been discussed in a variety of contexts in ethical thinking, most prominently by the philosopher Joel Feinberg (1984), who describes harms of this kind as *accumulative harms*. Andrew Kernohan (1995) summarizes accumulative harms as "the kind of harm brought about by the actions of a group of people, when the actions of no single member of that group can be determined to have caused the harm, and when, as far as we know, no single action taken by itself has enough impact or is likely to have enough impact to be called harmful." This concept of accumulative harm raises serious questions about understanding the problem of atmospheric pollution in terms of violating rights, due to the unresolved question about who is *causing* harm and who has duties to the rights holder.

Another way to understand the ethical questions surrounding atmospheric pollution is to consider the suffering it can cause, both to human beings and to other animals—an important issue for those who regard the minimization of suffering as a fundamental ethical principle. However, atmospheric pollution is a side effect of activities that in themselves reduce suffering and increase happiness. So it is important in this instance to weigh the costs and benefits against one another. But to do so raises issues of justice, since it has been frequently observed that the greatest producers of pollution are seldom those who suffer the worst harm (families too poor to own cars live in housing on polluted urban streets, while the car drivers live in less polluted suburbs). This justice problem is even more acute when we consider animals, who may feel the effects of air pollution but very rarely gain any of the benefits.

Neither of these perspectives takes into account the more general *environmental* harm caused by atmospheric pollution: damage to individual living organisms, ecosystems (such as forests or lakes affected by acid rain), and, potentially, species. For some ethicists, these harms merely add to the more direct harms caused to people and animals by air pollution, primarily because they reduce crop yields and destroy esthetically pleasing areas of forest. But for others, the environmental damage caused by atmospheric pollution is of direct ethical concern. These ethicists maintain that all individual living things, whether humans, animals, plants, or bacteria, have value in themselves, independent of their uses to human beings. If an organism is killed by acid rain or increased exposure to ultraviolet light, value is lost to the world, regardless of whether human beings have a use for the organism, or even notice its death. Some ethicists also maintain that ecosystems and species are valuable as wholes, whether or not they are useful to human beings (see entries on Ecosystems and Biodiversity). Therefore, damage to ecosystems and extinction of species is, from this perspective, ethically wrong.

It is indisputable that atmospheric pollution causes a variety of harms to human beings, other organisms, and the environment. A variety of attempts to mitigate such harms have been proposed (such as the banning of the production of CFCs, the installation of pollution control equipment on power stations, etc.). For some ethicists, these measures go a good way toward relieving the problem of atmospheric pollution; they hope that technological changes will reduce such problems even further. For other philosophers—in particular those who identify themselves with the deep ecology movement—such technological changes are insufficient to come to terms with the kind of environmental and social problems created by atmospheric pollution. This group of ethicists advocates a change in the nature of technological society, so that, for instance, less energy is consumed, and the energy that is used comes from nonpolluting sources. Atmospheric pollution, then, raises a number of fundamental questions about our value systems and the kind of lifestyles we have chosen.

References: Feinberg, Joel. 1984. *The Moral Limits of the Criminal Law*, vol. 1: *Harm to Others*. Oxford: Oxford University Press.

Kernohan, Andrew. 1995. "Rights against Polluters." *Environmental Ethics* 17(3): 245–259.

Oden, Svante. 1968. "The Acidification of Air and Precipitation and Its Consequences on the Natural Environment." Ecology Committee Bulletin No. 1. Stockholm: Swedish National Research Council.

Organization for Economic Cooperation and Development.. *The State of the Environment, 1991.* Paris: OECD.

United Nations Environmental Programme. 1993. *The World Environment 1972–1992.* London: UNEP/Chapman and Hall.

Biodiversity

The word *biodiversity* is an abbreviation for biological diversity—the existence of a wide variety of kinds of life on earth. It includes diversity of individuals, diversity of genes, and diversity of species.

Biodiversity can be measured in different ways: with reference to the genetic variety within members of a species (practically, this information is very difficult to ascertain), with reference to the variety of species in existence or *species richness*, and with reference to the variety of communities or habitats created by different species (Pellew, 1995). Most commonly, however, the second of these alternatives, species richness, is used to measure biodiversity, both in a particular area and globally. It is reasonable to assume, then, that a rise in the number of species in the world means a rise in biodiversity and a fall in the number of species in the world means a fall in biodiversity; however, even this definition can be problematic.

Also, it is important to note that the whole idea of species is a much contested and debated construction among biologists. The most common definition of a species is a group of organisms whose members can interbreed and produce fertile offspring. But classification of species has not always reflected this definition, since members of some separate species can interbreed and produce fertile offspring. This problem could ultimately lead to reclassification of species, which could give the appearance of a rise or fall in biodiversity that does not reflect any change in the natural world.

Partly because of the disagreement about classification, but mainly because of the magnitude of the task, there is no agreement about the number of species currently in existence on earth. Estimates vary between 5 million and 100 million, with 10 million to 30 million the most commonly cited figures. Of these species, about 1.5 million have been formally identified. Throughout the evolution of life on earth, species have evolved and become extinct. Current estimates put this "background level" of extinction at one species per million in existence per year. There is general concern that at present, owing to human activity, species are becoming extinct at a much higher rate, perhaps 100 to 10,000 times the background rates, although some biologists, such as Julian Simon, dissent from this view (Fletcher 1995). The main reason for such species loss appears to be the destruction of habitat through development (such as the draining of swamps and deforestation), although pollution and hunting and fishing have been contributing factors.

Biodiversity varies across different parts of the planet. On land, areas of tropical rain forest tend to have the highest intensity of biodiversity, measured in terms of species richness. Isolated islands tend to have the greatest number of species unique to them (endemic). Therefore, the destruction of these two areas has the greatest impact in terms of loss of global biodiversity.

National and International Law and Policy A number of national and international agreements exist to protect biodiversity. Nationally, the most important legislation is the Endangered Species Act (ESA) of 1973 (with subsequent amendments). Under the ESA, species can be listed either as endangered or threatened; once they have been listed, they and their habitats are entitled to legal protection. This protection applies not only on federal land but also on land owned privately. In some cases private landowners can be prevented from developing their land—one of the controversial areas of the ESA. Currently, 1,520 species of plants and animals are listed under the ESA, 965 in the United States and its territories (Corn 1996). Listed species include the spotted owls of the Pacific Northwest and sea turtles in the Gulf of Mexico.

Internationally, there are two significant agreements aimed at the protection of biodiversity. The oldest of these is the Convention on International Trade in Endangered Species of Wild Flora and Fauna (CITES). This convention came into force for signatories in 1975 and has, as of 1994, 122 signatories. The United States signed CITES in 1973, and the ESA represents, in part, the incorporation of CITES into U.S. law. Thus, violations of CITES on U.S. soil are violations of the ESA. It is worth noting, however, that CITES focuses on international *trade* in endangered species and parts of endangered species. It may therefore be applied to cases such as the smuggling of ivory or tiger products for medicine. However, since most extinctions are due to habitat loss rather than trade, CITES cannot fully address the problem of species extinction.

A second international agreement, the Convention on Biological Diversity, was signed at the United Nations Conference on Environment and Development (UNCED) held in Rio de Janeiro in June 1992. Initially, President George Bush refused to sign the treaty, expressing concern about pledges in the treaty concerning intellectual property rights and offers of financial assistance to developing countries. President Bill Clinton signed the treaty in 1993. The treaty affirms that the conservation of biodiversity is a "common concern of mankind," but it also maintains that nations have "sovereign rights" over their own resources. This convention establishes three major goals: to conserve biodiversity; to use its components sustainably; and to distribute the benefits derived from genetic resources equitably. Its 42 legally binding articles include the development of national strategies to conserve biodiversity, regular meetings to review compliance and consider further actions, the establishment of protected areas where special measures are needed and the sustainable management of these and surrounding areas, legislation to protect endangered species, education about and research into issues concerning biodiversity, safety in the release of genetically engineered organisms and the introduction of alien species into existing ecosystems, and transfer of technology that would promote biodiversity.

Relevant Ethical Issues A variety of ethical issues is raised by biodiversity and species extinction, and the subject has been one of considerable interest to environmental ethicists. A number of central, interlocking questions have been raised: Is maintaining biodiversity—or aiming to increase it—a good thing? Do we have an ethical duty to preserve species? Is the extinction of species—a process that has been occurring since life evolved on earth—necessarily a bad thing?

The most common answer to these questions is that biodiversity is ethically significant because it is extremely—maybe vitally—useful for human beings and should therefore be protected as a valuable resource. By protecting biodiversity, we preserve a storehouse of genetic material for possible use in the future—material that might, for instance, prove useful in genetically engineering new crop strains. Many species may also have a vast range of medical and industrial uses that are as yet unknown. It is also sometimes argued that biodiversity is essential for the maintenance of ecosystems and the "services" they provide human beings by recycling water, air, and nutrients. The loss of biodiversity thus might result in ecosystem collapse and problems for future generations of human beings.

These "human resource" answers, however, provide only weak arguments for the protection of biodiversity. They are fundamentally based on human ignorance: because we don't know what benefit any particular species may have for human beings, we had better preserve them all. Such arguments are at best provisional. Research suggests that only a very small number of species would be useful for crop engineering, for instance; samples of likely plants grown in collections should suffice (Pellew 1995). Similarly, it has been argued that it may not be biodiversity itself that is essential for the provision of "biological services" such as nutrient and water recycling, but perhaps just biomass; a monoculture may, in some cases, perform the task just as efficiently. Thus, such human resource arguments do not provide strong ethical reasons why biodiversity should be protected.

Other human-focused arguments for the protection of biodiversity are also proposed. It is argued that species have great esthetic and cultural value for human beings and that the loss of such values is an ethical problem. However, although this might be true of some species, it is clearly a selective argument. Many people may have mourned the loss of the last passenger pigeon, or regretted that they have never seen a dodo. However, such regrets rarely extend into the insect or bacterial realm and might not even apply to all mammals, and, in many cases, species become extinct before they have been identified, let alone become culturally or esthetically important. Again, this approach does not provide a strong argument for the protection of bio-diversity in general, but only of particular cherished species.

These arguments by no means exhaust ethical debate about biodiversity. Much debate has focused on the question of intrinsic value—whether species might be valuable in themselves, regardless of their possible uses for human beings. This question has, of course, profound implications for the discussion about conserving biodiversity (measured as species richness). In popular ethical writing, it is quite common to read the statement that "species have rights." Certainly, it would be legally possible to write legislation that gives species rights (and some have argued that this is precisely the effect, if not the actual wording, of the Endangered Species Act). But it is hard to find a philosophical explanation of where such rights might come from. The idea of rights is controversial in philosophy, and many philosophers do not accept that rights can be anything except political constructions. Other philosophers have argued that rights can be based on membership in the human community or on the complex individuality of human beings. Neither of these explanations can easily be extended to include species. So environmental ethicists have largely focused on explanations other than rights as to why species might be valuable independent of human use. The philosopher Holmes Rolston, for instance, argues that species are "living historical forms." Extinction of a species not only kills the individual, it kills the whole form, all future possible individuals of that type. Seen from this perspective, individual organisms are merely tokens of their type, of less significance than the type or species itself: "The life that the individual has is something passing through the individual as much as something it intrinsically possesses. The individual is subordinate to the species, not the other way round" (Rolston 1988).

Lawrence Johnson (1992) makes a rather different argument in favor of the noninstrumental value of species. He maintains that the "organic unity" of a species is such that it can be described as a kind of "super-organism," displaying many of the qualities of an ordinary organism, such as the ability to reproduce itself. From this premise, he argues that species can have interests, in the same way other organisms have interests (for instance, it is in the interests of a houseplant to be watered and a dog to be fed). Thus, just as we might accept that we have an ethical duty to meet the dog's interests by feeding it (to keep it alive), so we might accept that it is an ethical duty to keep a species alive, by protecting it.

Clearly, there are a number of problems with both of these arguments. Rolston's position assumes the existence of values in nature that are not created or bestowed by human beings; this is in itself a contentious argument. Johnson's position assumes that a species is the same kind of thing as an organism and that we have moral duties toward all kinds of organisms. However, both of these assumptions are often disputed. It is hard to maintain, then, that species *as species* (rather than as their individual members) are valuable independent of human use.

This conclusion does not mean that all ethical arguments for the protection of species are misguided. The consequences for both present and future human beings of the widespread loss of biodiversity are widely acknowledged as significant. Furthermore, the consequences for living organisms in general—both for individual members of threatened species and for those individuals living in an ecosystem from which a key species has become extinct—may be very serious. In most practical cases, a variety of ethical reasons can be provided for protecting biodiversity.

References: Corn, Lynne. 1996. "Endangered Species: Continuing Controversy." Congressional Research Service Report for Congress, March 18, 1B 95003.

Fletcher, Susan. 1995. "Biological Diversity: Issues Related to the Convention on Biodiversity." Congressional Research Service Report for Congress, May 15, 95-598 ENR.

Johnson, Lawrence. 1992. *A Morally Deep World*. Cambridge: Cambridge University Press.

Pellew, Robin. 1995. "Biodiversity Conservation: Why All the Fuss?" *RSA Journal* (January–February): 53–65

Rolston, Holmes. 1988. *Environmental Ethics: Duties to and Values in the Natural World*. Philadelphia: Temple University Press.

———. 1994. *Conserving Natural Value*. New York: Columbia University Press.

Climate Change

In its most general sense, climate change refers to long-term changes in weather patterns (including air and sea temperature, wind speed and direction, precipitation, and air pressure), usually (but not always) on a global scale, whether or not these changes are of human origin. Most commonly, the term is now used in a more particular sense to describe the changes in global climate that may be caused by increased anthropogenic emissions of the gases that increase global warming.

The earth's climate has always been warmed by the presence of gases in the atmosphere that absorb solar radiation, trapping the sun's heat by preventing it from being reradiated away from the earth in much the same way a greenhouse traps heat. Hence, these gases are often called greenhouse gases. This natural effect has maintained a global mean surface temperature about 33°C higher than would be the case otherwise (UNEP 1993). Human industrial and agricultural activity, especially in the last 150 years, has released substantial quantities of these greenhouse gases (especially carbon dioxide, methane, nitrous oxide, and chlorofluorocarbons) into the atmosphere. During the early 1970s, scientific concern began to grow that these additional emissions of greenhouse gases would enhance the natural process of global warming, leading to higher global temperatures and corresponding climate change. This concern increased during the 1980s and 1990s as computer climate models predicted enhanced global warming. In particular, measurements of global temperatures indicated that a significantly high number of the warmest years this century occurred during the 1980s and 1990s. Some surface data suggest that 1995 was the warmest year since the historical climate record began (Justus and Morrissey 1996).

The specific effects of such climate change at a local level are uncertain, but climate scientists have predicted rising sea levels (leading to the flooding of low-lying areas), increasing numbers of violent weather events such as hurricanes, desertification in some areas, and the thawing of permafrost in others. Such changes could have a significant effect on world agricultural patterns, on global forestry, on human health (for instance, through the wider spread of tropical diseases such as malaria), and could lead to the destabilization of some cultures and the creation of environmental refugees. However, although the majority of scientists researching this issue agree that some degree of enhanced global warming is occurring and will continue to occur, this is still a contested area of science. A minority of scientists argue that other global mechanisms (such as stratospheric ozone depletion and an increase in atmospheric sulfates) will counteract the effects of increased emissions of greenhouse gases.

International Policy on Climate Change Concerns about climate change were first expressed in the international political arena at the Stockholm Conference in 1972, which led to a series of international conferences on climate change. In 1988, the Intergovernmental Panel on Climate Change (IPCC) was formed by the World Meteorological Organization and the United Nations Environment Program. This panel was established to investigate human-induced climate change and to provide independent scientific advice. It created three working groups: one focusing on the science of enhanced global warming; one on the possible environmental impacts of such global warming; and one on possible responses to climate change, including economic and political issues. The panel produced its first full Assessment Report for the 1990 World Climate Conference, which was updated in 1992. It forecast an increase in global temperatures of between 1.5°C and 4.5°C by 2050 (U.K. Department of Environment 1994).

In response to this report, under the auspices of the United Nations, work began on an international convention on climate change, which was opened for signature during the 1992 Rio Earth Summit. The aim of this agreement is to stabilize "greenhouse gas concentrations in the atmosphere at a level which would prevent dangerous anthropogenic interference with the climate system." This agreement has now been signed by more than 160 countries, committing them to create national programs to limit their release of greenhouse gases and to create "sinks" for carbon dioxide (such as by afforestation programs); to take measures aimed at reducing their emissions of carbon dioxide in the year 2000 to 1990 levels; to help developing countries combat climate change; and to make provision for further rounds of negotiation relating to the period after the year 2000. In response to this agreement, the United States produced its own Climate Change Action Plan in 1993, which presents a suite of voluntary actions to reduce carbon dioxide emissions, focusing in particular on improvements in technology to increase energy efficiency and on afforestation programs (Justus and Morrissey 1996).

Further reports of the IPCC since 1992 have continued to warn of the likelihood and potential social and economic impact of global warming and have continued to recommend a cut of 60 percent in global emissions of carbon dioxide. It seems extremely unlikely that such cuts will occur, since many countries are unlikely to meet even the currently agreed target of returning emissions in 2000 to 1990 levels, and developing countries, especially China, are rapidly increasing their own carbon dioxide emissions.

Relevant Ethical Issues A wide range of ethical—and specifically environmental ethical—issues are raised by anthropogenic climate change. These are rendered particularly difficult and complex by the contested and uncertain nature of the scientific work that surrounds the issue. In particular, climate change raises questions about what to do ethically in situations where the degree of the risk and the nature of its consequences are unknown.

If we take a worst-case scenario, then enhanced global warming could have a variety of negative effects on present and future human beings. It could create millions of environmental refugees suffering hunger and homelessness; reduce agricultural output, resulting in possible famines; create and aggravate water shortages in some places while leading to floods in others; lead to an increased human death toll in violent weather episodes; extend tropical diseases over much wider areas of the planet; and lead to a difficult future for people yet to be born. Nearly all approaches to ethics would acknowledge some or all of these results to be ethically undesirable. Environmentally, enhanced global warming could lead to a reduction in biodiversity with a significant increase in species extinction, the loss of a variety of rare ecosystems, suffering and/or death to millions of organisms, and the destabilization of global climate systems with potentially serious consequences for all life on earth. Most environmental ethicists would view some or all of these outcomes as ethically unacceptable.

Were such effects certain, the ethical imperative to act would be a very strong one (although even this would be no guarantee of political action; practices known to be harmful have not always been prevented politically). However, where the science is uncertain and reducing emissions of greenhouse gases creates ethical problems of its own, whether to act and what kind of action to take is more problematic. Cutting emissions of carbon dioxide is especially difficult, as most emissions come from the burning of coal and oil, currently the engines of economic development. Reducing carbon dioxide emissions in developed countries by 60 percent to lessen the risk from enhanced global warming, as recommended by IPCC scientists, would require a reorientation of developed economies toward generally lower economic activity, and in particular a substantial reduction in the use of private motor vehicles. Yet this in itself may have a social cost in terms of higher unemployment, loss of personal freedom and convenience, and lower levels of prosperity. Such consequences also raise ethical questions, although these primarily relate to human beings rather than the environment. (It is hard to see how the natural environment, or elements of it, could be harmed by a slowdown in economic growth of this kind, unless energy generation were switched to other potentially environmentally destructive methods such as hydroelectricity or nuclear power.) Because of this uncertainty, some governments—including the U.S. government—have adopted what is known as a "no regrets" policy, where the steps taken to reduce the emissions of greenhouse gases must also benefit (rather than harm) the human community. Thus, U.S. policy focuses on practices such as tree planting and technological improvements that can produce benefits besides reducing emissions of greenhouse gases.

But given the possibility of the worst-case scenario proposed above, is this no regrets policy an adequate ethical response? No regrets measures come nowhere near producing the kinds of reduction in greenhouse gases many scientists consider necessary to avert significant climate change. Could it not be argued that the sacrifice of some present goods (in terms, for instance, of jobs or transportation preferences) should be accepted in order to protect the future of human beings, other organisms, ecosystems, and species, even where the risk is known? This response would certainly be suggested by the *precautionary principle*, a principle now fundamental to global environmental policy. Although definitions of this principle vary, the precautionary principle generally prescribes that in situations of scientific uncertainty where there is risk of serious environmental harm, policy should err on the side of caution. If such an approach toward global warming were seriously adopted, we would expect much more stringent attempts to reduce outputs of greenhouse gases.

Other ethical approaches may also be taken toward climate change. Many deep ecologists, for instance, already reject the idea of a fossil fuel–driven economy for philosophical and spiritual reasons that have nothing to do with climate change. They argue that climate change is a *symptom* of the problem created by industrialized lifestyles that are disharmonious with nature and treat the natural world as a resource to be exploited. Rather than focusing on curing the *symptom*, we should focus on curing the *disease* by radically changing lifestyles and values—which would have the effect of preventing global warming (see, for instance, Lemons 1983; Devall and Sessions 1985). Other kinds of ethicists, influenced by Lovelock's Gaia hypothesis (although Lovelock himself does not argue this), point out that life on earth in some form has survived dramatic climate change before and can do so again. In the process, the human species and some other forms of life may be lost, but is this really an *environmental* ethics problem? Isn't it primarily a human ethics problem? From a global "Gaian" perspective, life on earth persists; it just takes different forms (*see* Gaia).

Human-induced climate change is one of the most threatening of global environmental issues. It is also the most complicated—scientifically, politically, economically, and ethically. Because the risks remain uncertain and a number of different human and environmental values are at stake, making clear ethical prescriptions is extremely difficult. It is probable, however, that if the effects of global warming become more obvious, then ethical choices will also become clearer; however, this does not make the implementation of present sacrifices for future goods any easier.

References: Devall, Bill, and George Sessions. 1985. *Deep Ecology.* Layton, UT: Peregrine Smith.
Global Climate Change. 1994. 3d ed. London: Department of the Environment/HMSO.
Justus, John, and Wayne Morrissey. 1996. "Global Climate Change." Congressional Research Report for Congress, 18 March, IB 89005.
Lemons, John. 1983. "Atmospheric Carbon Dioxide: Environmental Ethics and Environmental Facts." *Environmental Ethics* 5(1): 21–33
United Nations Environmental Programme. 1993. *The World Environment 1972–1992.* London: UNEP/Chapman and Hall.

Deforestation

To deforest is to fell forests, to deprive of forests (usually understood to mean without replanting, or felling at a greater rate than the rate of replanting).

Deforestation has been increasingly viewed as a global problem. Although it has stabilized in the Northern Hemisphere and in temperate areas, deforestation in tropical areas is still increasing. On the basis of satellite surveillance and ground checks, the Food and Agricultural Organization (FAO) of the United Nations estimates that whereas 11 million hectares of tropical forest were felled during 1981, 17 million hectares were felled during 1990. Causes of deforestation in tropical areas vary. Only a small percentage is felled for timber production (either within the nation or as part of an international timber market). However, this small amount may have disproportionate effects if logging roads are built into the tropical forests. Logging roads open up forests to agriculture, and agriculture is the most common reason for tropical deforestation.

In the United States, recent debates about deforestation have focused on the felling of old-growth forests, primarily in the Pacific Northwest. Old-growth forests are forests dominated by trees over a certain age (age is species-dependent) and often means "natural" forests. These forests provide high-quality timber, and felled areas can be replanted as managed forests that can be harvested after 60 to 100 years (Booth 1994). To date, about 85 percent of these old-growth forests in the United States have been felled and replanted with timber plantations.

Deforestation and International Law and Policy   A variety of organizations and agreements govern the management of the world's forests. Most important among these are the International Tropical Timber Organization (ITTO) founded by the International Tropical Timber Agreement, the Tropical Forestry Action Programme (TFAP), and the Forest Principles, which stem from the 1992 Rio Earth Summit.

The International Tropical Timber Agreement formed the basis of the ITTO in 1983. Although the organization's aim was to promote the export of timber from tropical forests, this agreement is the only binding international commodity agreement dealing with the management and conservation of tropical forests (Lyke and Fletcher 1992). It aims to strike a balance between the interests of conservation and the sustainable use of tropical forests. Revised in 1994, the agreement aims to achieve sustainable management of all tropical forests by 2000 (Fletcher 1995).

The Tropical Forestry Action Programme was launched in June 1985. It was intended to help development assistance agencies coordinate funding of national and regional sustainable forest management plans. However, the TFAP was criticized by environmental pressure groups for actually encouraging rather than restraining deforestation. In 1990, the TFAP was reformulated with the specific aim of curbing tropical forest loss and promoting sustainable use of tropical forest resources.

Deforestation was one of the key topics on the agenda at the 1992 Rio Earth Summit. However, the two documents concerning forests that emerged from the summit are widely regarded as weak and lacking in authority. One of these documents, the Forest Principles, is a "non-legally binding authoritative statement of principles for a global consensus on the management, conservation and sustainable development of all types of forest" (Lyke and Fletcher 1992). As non-legally binding principles, they have little power to prevent deforestation; environmentalists argue that the revised TFAP is actually a stronger statement of forest principles. The other document is a chapter in a much larger document, Agenda 21, entitled "Combating Deforestation." It considers forest management under a variety of headings—protecting, conserving, sustainably managing, efficiently utilizing—but also lacks legal force. Thus, the protection and sustainable management of forests globally have little support in international law.

Forests have continued to be a focus of international environmental debate since 1992. In 1995, the U.N. Commission on Sustainable Development established an ad hoc Intergovernmental Panel on Forests. The task of this panel is to examine five areas: scientific research, trade, aid, the role of international institutions, and the implementation of the 1992 agreement on forests. It will report to the U.N. Commission on Sustainable Development in 1997 (Humphreys 1996).

Relevant Ethical Issues   Deforestation raises a wide variety of ethical issues. Many of these concern the loss of human opportunities brought about by deforestation, in particular the loss of home and livelihood for forest dwellers and the destruction of recreational areas and areas that may be of scientific interest. Alongside these human opportunity costs, deforestation also causes a wide variety of environmental problems, and for this reason it is of central concern to environmental ethicists. In summary, deforestation raises ethical questions about the loss of biodiversity and unique ecosystems as well as the loss of life and suffering of individual organisms. Additionally, deforestation in the tropics (because such areas have poor soil and are rarely reforested) may contribute to soil erosion and local and global climatic problems.

First, then, and most obviously, forests can be high in biodiversity. Tropical forests in particular are areas both of high biodiversity and high endemism (species unique to that particular environment). A small patch of tropical forest may contain 40,000 different insect species (Palmer 1992). The felling of tropical forests results in a loss of biodiversity and species extinction, the extent of which is unknown since so few species in tropical forests have been identified. This loss of biodiversity is a particular concern of ethicists, both because biodiversity may be valuable to human beings and because it may have intrinsic value apart from human use (*see* Biodiversity). Ethical questions can also be raised about the loss of cultural diversity resulting from deforestation. Many indigenous peoples still live in tropical forests and follow cultural practices established over thousands of years. Deforestation can force these peoples to leave forests or to have increased contact with alien cultures, which ultimately results in the loss of their own cultural uniqueness. Although some ethicists do not consider this cultural loss to be problematic, others argue that cultural diversity is as valuable to human beings as natural diversity.

Alongside the loss of natural and cultural diversity that may result from deforestation are the effects it may have on individuals—of all kinds. Clearly, human death and physical and psychological suffering can result from deforestation in areas inhabited by indigenous forest peoples (not only from loss of homes and cultural traditions but also from exposure to alien diseases). Deforestation also causes death and suffering to animals in the forest, both during the process of felling the trees and through the resulting loss of habitat. For many animal ethicists concerned with animal well-being and the reduction of suffering (such as Peter Singer), deforestation thus causes ethical problems. Some ethicists argue that the wrong of deforestation is not confined to the suffering of sentient animals. Paul Taylor argues that humans should show respect for all life-forms, whether or not they are capable of suffering. He thus argues that even the felling of a tree is the taking of an individual life and is ethically wrong. The felling of a forest, with the loss of huge numbers of individuals of all kinds, is thus of intense ethical concern.

A further ethical concern generated by deforestation is the loss of the ecosystem as a whole, not just as a reservoir of species and individuals. For some ethicists, the ecosystem is the primary location of ecological value.

Individual members are only important inasmuch as they assist in the working of the whole (*see* Ecosystems). While such an ethical approach may allow for selective logging in forested areas, deforestation understood as the clear-cutting of forests means the loss of whole, valuable biological communities. Such destruction is irreversible, even by reforestation projects, because "old-growth forests have unique structural features and provide unique wildlife habitats, unlike those available in managed forests" (Booth 1994). From this perspective, deforestation is particularly ethically problematic.

Forests also have effects on local and global climates. They are vital to local cycles (such as the hydrological cycle and the nitrogen cycle). Loss of forests (even with replanting) can cause disruption of these cycles, causing harm to human beings and other organisms and ecosystems. Forests are also important in stabilizing global climate. Trees act as sinks or stores for carbon dioxide, thus preventing substantial quantities of one of the main gases implicated in enhanced global warming from entering the atmosphere (*see* Climate Change). Deforestation—especially where forests are burned—results in the release of stored carbon dioxide into the earth's atmosphere and can add to global warming. It may thus have problematic consequences for both present and future human beings and for other living organisms.

Deforestation, then, raises complex ethical issues and causes a variety of different losses and harms. Yet, as is often the case with complex ethical questions, ending deforestation might also cause harms. Thus, some ethical arguments, from a human perspective at least, support continued deforestation. Alistair Gunn (1994) points out that "if logging were to stop immediately, the economies of Kalimantan and Sulawesi in Indonesia and Sarawak and Sabah in Malaysia would collapse." Deforestation provides some developing countries with crucial income and provides individual humans throughout the world with work and places to grow food. The setting aside of untouchable forest preserves may have the effect of creating areas of high unemployment and in tropical forests may drive indigenous peoples from their homes, making forest areas the domain of rich foreign tourists. Furthermore, southern governments may well argue that it is not their deforestation that is contributing so substantially to enhanced global warming, but rather the high energy consumption of rich northern economies. Many humans, including some of the very poor, have a considerable amount to gain from continued deforestation. However, deforestation also causes suffering and death to a variety of organisms (sometimes including humans), destroys ecosystems and biodiversity, and exacerbates global warming.

Perhaps the argument holding the trump card is the finite nature of tropical and old-growth forests; they are irreplaceable. When they are gone, the problems of agricultural land and unemployment will return, but there will be no more forests to stave them off. If this is the case, it can be argued that these problems should be addressed now while some forests still exist. Other things being equal, most environmental ethical arguments suggest that a world with the same amount of poverty and unemployment but with tropical forests is of more value than one without.

References: Booth, Douglas. 1994. "The Economics and Ethics of Old Growth Forests." *Environmental Ethics* 14(1): 1992.

Fletcher, Susan. 1995. "International Forest Agreements: Current Status." Congressional Research Service Report for Congress, 11 September, 95-960 ENR.

Forest Resources Assessment 1990 Project, Food and Agriculture Organization of the United Nations. "Second Interim Report on the State of Tropical Forests." Paper presented at the Tenth World Forestry Congress, Paris, September 1991.

Gunn, Alistair. 1994. "Environmental Ethics and Tropical Rain Forests: Should Greens Have Standing?" *Environmental Ethics* 16(1): 21–41.

Humphreys, David. 1996. "The Global Politics of Forest Conservation since the UNCED." *Environmental Politics* 5(2): 231–258.

Lyke, Julie, and Susan Fletcher. 1992. "Deforestation: An Overview of Global Programs and Agreements." Congressional Research Service Report for Congress, 21 October, 92-764 ENR.

Palmer, Joy. 1992. "Destruction of the Rainforests." In David Cooper and Joy Palmer, *The Environment in Question*. London: Routledge.

Singer, Peter. 1973. *Animal Liberation*. 1983 ed. Wellingborough, England: Thorsons Press.

Taylor, Paul. 1986. *Respect for Nature*. Princeton, NJ: Princeton University Press.

Ecosystems

What might be meant by the term *ecosystem* is widely debated by ecologists and environmental ethicists. This debate over definition is inseparable from the broader discussion about the existence and nature of ecosystems, since the term is impossible to define without making some judgment about the nature of an ecosystem itself. A standard dictionary might define an ecosystem as "a unit consisting of a community of organisms and their environment." However, the word *community* replaces one problematic concept by another, while the idea that an ecosystem is a "unit" is also a contested issue. It is therefore better to sidestep the definition itself and instead consider the debate about the nature of ecosystems.

What Is an Ecosystem?   The dominant view among ecologists has varied during the twentieth century. As early as 1916, the ecologist F. E. Clements argued that an ecosystem was a kind of organism or superorganism that persisted over time and in which all the elements were closely integrated with and dependent on one another, in the same way as the different parts of organisms (such as human beings) are related to one another. Though few ecologists accepted this view in as extreme a fashion as Clements, the term *ecosystem* was widely used in ecology during the 1960s to describe a spatially limited, closely knit group of organisms in a particular environment. However, alongside the ecologists who affirmed the closeness—even organic closeness—of organisms in an ecosystem were ecologists who disputed this. As early as 1952 the biologist H. A. Gleason commented, "Far from being an organism, an association is merely the fortuitous juxtaposition of plants." More recently, ecology seems to have moved away from talking in terms of ecosystems at all. Some ecologists argue that since it is impossible to draw boundaries between different ecosystems, and since the degree of wholeness and integration ecosystems display is questionable, the whole concept of an ecosystem should be abandoned. Some ecologists now prefer to understand ecosystems in terms of dynamic, interrelated processes over time. (See Callicott 1995.) Others suggest that ecosystems do contain some degree of internal integration, although they are loosely organized (Rapport 1995).

Relevant Ethical Issues The lack of agreement about the nature of ecosystems lies at the base of a number of problems in environmental ethics, partly because some—but by no means all—environmental ethicists want to affirm the value of groups or collectives rather than just individual organisms. (After all, the term *environment* is not usually used to describe individuals.) One kind of group that some environmental ethicists argue is valuable is the ecosystem (see, for instance, the entry on Deforestation). However, if it is erroneous to talk about ecosystems, then it makes no sense to argue that they are valuable.

This point raises a whole raft of difficulties in environmental ethics. First, there is the question of whether parts of the natural world might properly be called ecosystems. Even if the existence of ecosystems of some kind is conceded (as dynamic, related processes, perhaps), other questions remain unresolved. For instance, what is it about an ecosystem that is valued? Is its very existence valuable, whatever condition it might be in? Or are there other things about it that are valuable—such as the degree of biodiversity it contains, or how healthy it is, or how stable and resistant to change it is? All of these ideas are deeply problematic; they rest on the premise that there are better and worse states for ecosystems to be in. But, for instance, is a forest ecosystem seriously affected by acid rain in a worse condition than one unaffected by it? What does "worse" mean here? An ecosystem, as Dale Jamieson (1995) points out, is hardly the kind of thing that could *mind* what condition it is in. Does "worse" mean, then, that human beings have particular states that they prefer ecosystems to be in?

If we accept that the diversity of ecological processes by which we are surrounded may be called ecosystems, then clearly human beings have good reasons for wanting them to persist in forms similar to the ones they have at present. Ecosystems are crucial to human survival, allowing for the continuance of natural cycles, especially of carbon and water, and assisting in climate regulation. Both presently living humans and future generations of humans are dependent on these free "ecosystem services" for survival. For these reasons alone, most ethicists (whether or not they accept the label "ecosystems") argue that such ecological processes are of important *instrumental* value to human beings, and that "better" states for ecosystems—from a human perspective—are states where they can carry out such functions effectively.

However, this is not all that many ethicists are getting at when they talk about valuing ecosystems or particular states of ecosystems. Some ethicists, for instance, argue that we can talk about ecosystem health independent of its usefulness to human beings. J. Baird Callicott (1995) argues that ecosystem health is "an objective condition of ecosystems" and that it is "prudently, aesthetically and intrinsically valuable." As such, he argues that achieving ecosystem health should be one of the goals of conservation. Working from very different principles, Holmes Rolston (1988) also argues that ecosystems are valuable irrespective of their usefulness to us; since human beings evolved through ecosystemic processes, it would be very odd to value the *product* without valuing the *process*. However, the views of both these ethicists are widely contested.

The debate in environmental ethics over the nature and value of ecosystems is a complex and difficult one, impossible to summarize or to resolve here. It is also a particularly interesting debate because it reveals how closely related environmental ethics and scientific ecology can be. Ethical positions based on outdated ecological understandings of ecosystems, for instance, cannot withstand critical scrutiny. But equally, value-loaded conservation decisions made by scientific ecologists look vulnerable unless they are underpinned by the painstaking work that has been carried out by environmental ethicists. This area is one of constant change and interchange and seems likely to be one of the central issues debated in environmental ethics in the future.

References: Callicott, J. Baird. 1995. "The Value of Ecosystem Health." *Environmental Values* 4(4): 345–362.

Clements, F. E. 1916. *Plant Succession: An Analysis of the Development of Vegetation.* Washington, DC: Carnegie Institute. Quoted in Callicott 1995.

Gleason, H. A. 1952. "Delving into the History of American Ecology." *Bulletin of the Ecological Society of America* 56: 7–10.

Jamieson, Dale. 1995. "Some Preventative Medicine." *Environmental Values* 4(4): 333–344.

Rapport, David. 1995. "More than a Metaphor?" *Environmental Values* 4(4): 287–310.

Rolston, Holmes. 1988. *Environmental Ethics: Duties to and Values in the Natural World.* Philadelphia: Temple University Press.

Energy Resources

Energy can be defined as the "ability to do work" (Peet 1992). *Energy resources* refers to the different forms of energy humans use to do work for them.

The world energy supply is made up of several kinds of energy derived from a number of sources: fossil fuel (coal, oil, and natural gas), nuclear, hydroelectric, geothermal, solar, wind, and wave. These energy sources can, broadly, be divided into two categories: nonrenewable sources (those that can be exhausted) such as coal and oil, and renewable sources (those that can produce energy indefinitely) such as solar, wind, and wave. At present, energy use is concentrated in nonrenewable sources of energy, primarily oil. In 1994, for instance, oil alone accounted for 40 percent of total energy consumed in the United States. Although there has been much dispute concerning the amount of oil that remains in the earth, most recent estimates suggest that 2,330 billion barrels of recoverable oil once existed, of which about one-third has now been extracted and consumed (Riva 1995). Estimates by the International Energy Agency suggest that the demand for oil will continue to increase, reaching 94 million barrels a day by 2010. This trend suggests that oil supplies might become scarce around the middle of the next century. Reserves of coal are much greater than those of oil, with an expected lifetime measured in centuries rather than decades.

Concerns about the exhaustion of fossil fuels have led to increasing interest in energy *efficiency* (where technical changes result in less energy being needed to do the same amount of work). It is estimated that in the United States, energy efficiency has resulted in a saving of $275 billion per year, more than half as much as the total energy cost of about $522 billion (Sissine 1996). Energy efficiency reduces the huge environmental impacts of energy production and consumption. These impacts include air pollution (causing acid rain and photochemical smog); the release of greenhouse gases from the use of fossil fuels; reduction in wildlife habitat and biodiversity through mining and dams; the disposal of toxic waste products including nuclear waste; esthetic losses caused by unsightly developments such as wind farms; and the poisoning of land by production residues (such as from gasworks).

A further issue of interest here is the idea of hard and soft energy paths, so-called by the physicist Amory Lovins. He divides energy technologies into "hard" technologies—centralized, capital intensive, large-scale, and characteristic of advanced societies—and "soft" technologies—decentralized, small-scale, less complex, and less intensive (Lovins 1977). Lovins claims (though his claims are contested) that the hard energy path, followed by all industrialized societies, is in the long-term economically impractical because it is too expensive to maintain. He therefore argues that, for economic reasons, industrialized societies should switch to soft energy technologies, combined with measures to improve energy efficiency.

Relevant Ethical Issues A variety of questions for environmental ethics surround the use of energy resources. These concern the choice of energy source and associated technologies as well as the amount of energy that is actually used.

Clearly, different combinations of energy sources and technologies create different environmental impacts. Nuclear power, for instance, generates nuclear waste and ionizing radiation and carries a risk of catastrophic accident. It may, therefore, have devastating environmental impacts in terms of causing human and nonhuman suffering and environmental harm. But coal-fired power stations emit sulfur dioxide, which can cause environmentally damaging acid rain; they produce large amounts of carbon dioxide, thus contributing to enhanced global warming (*see* Climate Change), and they use up a nonrenewable resource. Hydroelectric dam projects, in their turn, divert rivers, killing river organisms and reducing river biodiversity, flooding huge areas, displacing and even killing animals (and, of course, people) and a variety of other living organisms. All these large-scale or hard methods of energy generation cause harm to other organisms and may reduce biodiversity and damage ecosystems. If such methods of energy generation are adopted, then value choices must be made about which of these environmental impacts is most significant. All of them have serious long- as well as short-term consequences (in terms of nuclear waste, climate change, or vast habitat alteration).

Because of the wide-ranging, long-term nature of the environmental harm caused by such technologies, some ethicists have urged that softer energy paths (such as the smaller-scale use of wind turbines and wave power) should be adopted. These technologies generally have lower environmental impacts (causing less harm to other living organisms and ecosystems); they often use renewable energy resources (such as wind, sun, and wave) and thus do not exhaust resources for future generations; and although they are not risk free, they generally have lower risks—economic, medical, legal, political, and environmental—associated with them (Shrader-Frechette 1984). However, questions arise about the ability of these softer energy pathways to provide the quantities of energy currently demanded by industrialized Western economies. Some of the shortfall, it is argued, may be made up by the pursuit of energy efficiency and energy conservation. From just about every ethical perspective, energy efficiency and energy conservation can be regarded as good things. If less energy is used, present humans benefit economically; future generations benefit by a lower consumption of nonrenewable resources; and humans, other living beings, and the environment suffer less short- and long-term harm from the impact of energy use. But even with energy efficiency and conservation measures, it is doubtful that soft energy pathways could provide sufficient energy to meet the energy needs of industrialized economies. This suggests that there may be ethical reasons for making a switch to a fundamentally less energy-intensive way of life, as well as switching between different sources of power and associated technologies to reduce their environmental impact.

This ethical issue, however, is much contested. Some policy makers argue that economic growth is dependent on increased energy consumption, and that without economic growth, nations will have no money to spend on tackling their environmental problems (an argument used, for instance, by the U.K. government in its environmental policy). For this reason, as long as environmental impacts are minimized, high energy consumption can be regarded—for environmental as well as social reasons—as ethically desirable. However, some environmentalists argue that economic growth is causing the environmental problems in the first place, so increasing economic growth can only make environmental problems worse. It is therefore ethically desirable to reduce both polluting energy generation and economic growth. Some elements of this dispute were vividly seen when, in the 1980s, two scientists claimed that they had discovered a chemical process that could produce atomic fusion and had the potential to generate huge quantities of cheap energy with little pollution or environmental impact. For many policy makers this discovery offered the possibility of continuing economic growth without the constraints and environmental damage caused by current energy generation. Environmentalists viewed the possibility of cheap, unlimited, and low-pollution energy as making economic growth—and hence environmental destruction—more likely more quickly. In any event, cold fusion does not seem likely to provide cheap, safe, and unlimited energy after all. But the fundamental nature of the ethical dispute between those who argue that more energy is needed and those who argue that we should reduce energy demand and live less energy-intense lifestyles is nonetheless highlighted. (See Sayre 1981 for a discussion of these differing value positions.)

Indeed, a variety of different views exists both on the *mechanisms* by which environmental protection may be achieved in the context of energy production and on how *benefits* to present human beings should be weighed in relation to future human beings, other animals, species, and ecosystems.

References: Lovins, Amory. 1977. *Soft Energy Paths*. New York: Harper and Row.

Peet, John. 1992. *Energy and the Ecological Economics of Sustainability*. Washington, DC: Island Press.

Riva, Joseph. 1995. "World Oil Production after the Year 2000: Business as Usual or Crisis?" Congressional Research Service Report to Congress, August 18, 35-925 CRS.

Sayre, Kenneth. 1981. "Morality, Energy and the Environment." *Environmental Ethics* 3(1): 5–19.

Shrader-Frechette, K. 1984. "Ethics and Energy." In Tom Regan, ed., *Earthbound: New Introductory Essays in Environmental Ethics*. New York: Random House.

Sissine, Fred. 1996. "Energy Efficiency: A New National Outlook?" Congressional Research Service Report to Congress, July 12, IB95085.

Gaia

In ancient Greek, the name *Gaia* refers to the goddess of the earth. In more recent times, the term *Gaia* has been used by an unconventional British scientist, James Lovelock, to describe a hypothesis about how the earth became and remains a place suitable for life. Lovelock first put this hypothesis forward in 1979 and expanded it in 1984. Since then he and others who have accepted his arguments have discussed the hypothesis and published articles about it in a range of forms, most notably perhaps in *Scientists on Gaia* (Schneider and Boston 1991).

In summary, Lovelock maintains in his Gaia hypothesis that all the constituents of earth, including living elements, oceans, atmosphere, and rocks, are part of a self-regulating living system that keeps the earth inhabitable for life. The evolution of life is coupled inseparably with the evolution of the physical surface of the earth and the chemical composition of the atmosphere. The so-called inanimate parts of the earth, Lovelock argues, are like the bark of a tree or the shell of a snail: an essential part of a living system. The atmosphere and temperature of the earth are kept, by the living organisms on it, as near to a constant (or in biological terms, to *homeostasis*) as possible. Where there are threats to this homeostasis—such as by the impact of huge asteroids that are often cited as the cause of several mass extinctions in the history of the planet—living organisms respond by a series of what Lovelock calls "feedback mechanisms."

As an example of such mechanisms at work, Lovelock points out that since the Proterozoic period between 1 and 2 billion years ago, the sun has become at least 30 percent and perhaps 50 percent hotter. Yet the temperature on earth has remained relatively constant and continues to support life. Lovelock argues that this relative constancy of the earth's temperature is caused by the feedback mechanisms operated by earth's living organisms. To illustrate how this feedback mechanism might work, in *The Ages of Gaia*, Lovelock created a computer model that he entitled Daisyworld. In its simplest form, Daisyworld is a planet rather like ours, except that only black and white daisies live on its surface. Lovelock assumes that these daisies can survive only between temperatures of 5°C and 40°C. As in our solar system, over millions of years, the heat of the sun slowly intensifies. Initially, when the sun is cooler, black daisies predominate. Being black, they can absorb the sun's heat and not only keep themselves alive but maintain the overall temperature of the planet at a level above 5°C. As the sun gets hotter, however, white daisies, by a process of natural selection, become increasingly common. With their ability to reflect sunlight, they can keep both themselves and the surface of the planet cooler and at a relatively constant temperature. Of course, there will still come a point where the sun's heat is such that, even with the entire planet covered with white daisies, they cannot reflect enough heat to keep the planet viable for life. At this point, the feedback mechanisms cease to function and Daisyworld becomes a deserted, barren planet. The effect of the living organisms, however, has been to keep the temperature of the planet between 5°C and 40°C for much longer than if the planet had been barren. (A barren planet would initially have been colder and would have reached 40°C much sooner.)

Although the earth is, of course, far more complex than Daisyworld, Lovelock contends that similar feedback mechanisms exist. Climate is regulated not only by the reflectivity of the land surface and of clouds, but also by the composition of the atmosphere—an increase in carbon dioxide, for instance, warms the earth by the so-called greenhouse effect (*see* Climate Change). Similar feedback mechanisms operate to keep the sea at constant salinity and oxygen at 21 percent of the atmosphere, a level high enough for fires (all-important in forest ecology) to occur, but not so high that any flame would kindle continental conflagration. For millions of years, then, the earth has thus responded to external changes such as asteroid impact and increased solar heat by processes involving the evolution not only of species but also of their environments. It is because of this ability of the earth to maintain conditions comfortable for life despite external threats that Lovelock sometimes refers to Gaia (the earth including biota, rocks, atmospheres, and seas) as a "living organism." Lovelock's hypothesis is disputed by many scientists, in particular those who think that it implies a "purpose" in evolution (something Lovelock himself has denied). The idea that the earth might in some sense be a self-regulating organism has also been attacked, in part on the grounds that the earth does not display the behavior of other organisms (respiration, excretion, and reproduction). However, there is no doubt that the Gaia hypothesis, in particular the idea of the earth as a living organism, has captured the imagination of many individuals concerned about the environment. Although its basis in modern science is new, the idea of the earth as a kind of living organism is not new with Lovelock; similar ideas were suggested in Plato's *Timaeus* and more recently underpinned the work of philosophers such as Georg Hegel and Alfred North Whitehead.

Relevant Ethical Issues   Although the Gaia hypothesis is popular among those who are concerned about environmental problems, it is not at all clear what its ethical implications might be. In fact, there have been several conflicting interpretations, few of which are acceptable to most environmental ethicists.

Lovelock himself makes one suggestion. Gaia's unconscious aim, he maintains, is to keep conditions on earth comfortable for the continuance of life on earth. In the earth's history, it has received various shattering blows, such as the impact of asteroids, which have caused substantial damage—"up to 60% burns," as Lovelock puts it. But the earth has recovered from these, just as any organism might recover from disease or accident, although it may have recovered to a new equilibrium, with new dominant species and even a different atmospheric chemistry. To destroy Gaia completely, rendering it a dead planet like Mars, would, according to Lovelock, be very difficult.

Such a position has significant implications for human beings. First, according to Lovelock, the earth is not particularly fragile; environmentalists who maintain that it is are deceiving themselves. Indeed, many of the concerns that occupy environmentalists—and environmental ethicists—today are, according to Lovelock, irrelevant to Gaia. The thinning of the ozone layer may increase human skin cancer, but Gaia has lived through times of much stronger solar radiation (Weston 1987). The dangers of nuclear power may terrify humans, but Gaia is largely indifferent. Even a catastrophic nuclear war, Lovelock argues, would by no means destroy Gaia; rather a new equilibrium would result, new species would evolve, and in a few hundred thousand years, Gaia would be as healthy as ever. Indeed, the only human behavior Lovelock thinks might threaten the survival of the living earth is the increase of greenhouse gases in the atmosphere, which could heat the earth beyond its ability to compensate with cooling feedback mechanisms.

With the possible exception, then, of global warming, human activities are not a threat to the life of the planet. The earth is either indifferent to them or will live through them but move to a new equilibrium in the process. The threat is rather to the survival of human beings (and to some species with which we are familiar—life as we know it), since they may well be eliminated in the process. In other words, Lovelock is concerned about human activities, not for Gaia's sake but for our own. If human beings continue to act in a way that disturbs Gaia's ability to maintain homeostasis, then Gaia may move to an equilibrium intolerable to humans, and the species may become extinct. This interpretation of the Gaia hypothesis would suggest that the environment is important in ethics not because *Gaia* is threatened with destruction but because it could destroy *us* if we are not careful in our behavior. With such conclusions, it is initially puzzling that the Gaia hypothesis has been championed by so many environmental groups. It contradicts the fundamental tenets of much environmental thinking about the delicacy and fragility of the planet. Indeed, it could even be used to support the view that whatever human beings do to the earth, with the exception of a large increase in greenhouse gases, the earth will respond with some kind of "clearing-up" feedback mechanism. However, it would be mistaken to suppose that this is the only way the Gaia hypothesis can be interpreted. The idea of a living earth is a potent symbol for many radical environmental groups, who maintain that the living earth is the foundation and location of value. With such underlying values, it can be argued that the health of the earth should be a human priority—if necessary, over and above the welfare of human beings, who are insignificant to the planet as a whole. Indeed, more radical adherents of this view have argued that human beings are like a cancer on the earth and that human diseases such as AIDS are a good thing, evidence of Gaia's feedback mechanisms swinging into action to protect life on earth. This turns Lovelock's ethical interpretation upside down, transforming the human-centered ethics that spring from his interpretation to a radical environmental approach to ethics.

Even these two opposing interpretations of the Gaia hypothesis do not exhaust the role it can play in environmental ethics. By far the greatest amount of material on Gaia is produced within what one might loosely call "green spirituality" approaches. Here, along with a host of other scientific theories, in particular quantum physics and scientific ecology, the Gaia hypothesis is used as evidence that "everything is interconnected" and is part of one living whole. The implication of this, it is claimed, is that the whole earth should be treated with reverence, as a living being. Sometimes, as in Theodore Roszak's classic environmentalist work *Person/Planet: The Creative Disintegration of Industrial Society* (1977), the earth is presented rather like a person, with human beings acting as its brain or nervous system. Such interpretations suggest the significance, even the vital nature of human presence on the living planet, but also maintain that it is possible for humans to live in harmony with the earth, in nonhierarchical, cooperative societies. What the ethical implications of this idea might be, in a world of human population growth and conflict over resources, is not made clear. But such an approach rejects both the entirely human-centered and the entirely earth-centered ethical approaches discussed above.

No unambiguous understanding of ethics flows from the Gaia hypothesis. It has been used to support a range of differing ethical positions. However, it has been important to many different kinds of environmental philosophy and should not be ignored in any study of environmental ethics.

References: Lovelock, James. 1979. *Gaia: A New Look at Life on Earth*. Oxford: Oxford University Press.
———. 1984. *The Ages of Gaia*. Oxford: Oxford University Press.
Roszak, Theodore. 1977. *Person/Planet: The Creative Disintegration of Industrial Society*. New York: Anchor Books.

Schneider, Stephen, and G. Boston, eds. 1991. *Scientists on Gaia*. Cambridge: Massachusetts Institute of Technology Press.

Weston, Anthony. 1987. "Forms of Gaian Ethics." *Environmental Ethics* 9(3): 217–230.

Genetic Engineering

The expression *genetic engineering* refers to the human technique of manipulating genes—units composed of deoxyribonucleic acid (DNA)—at the level of the cell or the molecule. The expression may be used as a general term to include other, similar processes, in particular recombinant DNA technology, where individual genes can be removed and made into molecular constructs (recombinant DNA) and stored, indefinitely if necessary (Walgate 1990).

The enormous potential for genetic engineering to change human life has been apparent since James D. Watson and Francis Crick discovered DNA in the 1950s and founded molecular genetics. Genes influence what a life-form will look like, what it can do, and issues such as what diseases a life-form is susceptible or resistant to. Genetic engineering provides humans with the opportunity to alter preexisting "blueprints" within life-forms by inserting, removing, or replacing particular genes. Such a procedure means that organisms can be fundamentally changed—made more or less resistant to disease, for example.

Genetic engineering has already had a substantial impact, especially on medicine and agriculture, and its impact will only increase. Genetically engineered material has been used in foodstuffs (where, for instance, vegetarian cheese is made with genetically engineered bacteria rather than rennet, an animal derivative), in animal experimentation (where mice have been genetically altered to be susceptible to cancer for experimental purposes), in medicine (where human genes have been inserted into sheep so that their milk contains Factor 9, a vital blood-clotting agent for hemophiliacs), in agriculture (where crops have been genetically engineered to be resistant to disease, drought, and herbicides), and in environmental science (where bacteria are being engineered to digest oil slicks and detoxify waste).

Genetic Engineering and National Law Genetic engineering raises a vast array of legal issues. Of interest here are those concerning animal welfare, release into the environment of genetically modified organisms, and the patenting of genetically modified material and organisms. Owing to the comparative newness of the technology, many legal matters are disputed and subject to review. In 1986, the U.S. government established a Coordinated Framework for the Regulation of Biotechnology, which clarified some of the legal issues surrounding biotechnology, though others remain unclear. In part, this lack of clarity is due to the number of U.S. agencies responsible for control of work in genetic engineering. The U.S. Department of Agriculture (USDA) is responsible for work on the regulation of plant pests and whole plants. The U.S. Environmental Protection Agency (EPA) is responsible for some areas of research under the Federal Insecticide, Fungicide and Rodenticide Act (FIFRA) and the Toxic Substances Control Act (although this legislation is not primarily designed to regulate research into, or the use of, genetically modified material). All field tests of genetically modified organisms must have an experimental use permit from the EPA. The Food and Drug Administration (FDA) has responsibility for genetic engineering where it relates to food safety, and in 1992 published regulations requiring companies working on genetically engineered plants to test for food safety and allergenic qualities and to address questions of nutrition. Some states (such as North Carolina in 1989) have introduced comprehensive legislation governing the release of genetically modified organisms into the environment.

But there is still concern over areas not currently covered. Little legal protection is offered, for instance, to animals being used for genetic experimentation. Livestock, mice, and rats are not covered by the Animal Welfare Act if they are being used in agricultural research, a category into which much genetic research falls. The National Institute of Health does have guidelines concerning the use of these animals in genetic experiments, but they apply only to projects funded by the National Institute of Health and are not usually monitored. A further area where no regulations exist is the labeling of food products produced by genetic modification; the genetically modified Flavr Savr tomato was approved by the FDA for marketing to the public with no requirement for labeling (Lee 1995).

Relevant Ethical Issues As is evident in the legal issues discussed above, many environmental ethics questions are raised by genetic engineering. These questions must include consideration of animals, not only because of the animal welfare issues raised by genetic engineering (although these are substantial), but also because of the underlying questions genetic engineering raises about human control and domination of life and living beings. Such questions are fundamental to the environmental debate, which is, in this sense, closely allied to the debate about genetic engineering.

At the most fundamental level, then, ethical questions are raised about the practice of genetic engineering itself. Such arguments can take several forms. Some commentators argue that genetic engineering is "unnatural," and that by manipulating genetic material humans are interfering in processes they should respect and/or revere. Such arguments are closely related to, and have a bearing on, arguments in environmental ethics that address the same questions from a macro rather than a micro perspective. They lead to fundamental and difficult questions about what might be meant by "natural" and "unnatural" and how far human behavior can be "unnatural." (Is modern medicine unnatural? Are modern farming techniques unnatural? Is the practice of animal domestication unnatural? Is homosexuality unnatural?) The debate about what natural means for human beings and whether the term has ethical implications is still hotly disputed (see, for instance, Norman 1996, Lucassen 1996). A minority of ethicists, however, reject all genetic engineering on the grounds of its unnaturalness (often on religious grounds of sacrilege).

A second fundamental ethical question about genetic engineering does not concern naturalness so much as the expression of control or domination over nature and living tissues that genetic engineering represents—an issue that is central to environmental ethics. It has been argued that genetic engineering is an extension of a technological mentality, often commented on in environmental ethics, which regards all living things as material available for human exploitation, a human resource. This issue is sometimes linked with a second, separate question: that of the patenting of genetic material. It is sometimes argued that the patenting of such material—essentially allowing life to be owned—intensifies this issue of domination. Peter Wheale and Ruth McNally (1990), for instance, suggest that "the extension of patent law to living organisms extends the exploitative, manipulative, and monopolistic worldview to the biosphere."

Other environmental ethicists maintain that particular values attach to genetic codes, either because they are a key constituent of the nature of individual organisms or because they identify entire species where species are regarded as valuable in themselves (see Biodiversity). For environmental ethicists who adhere to such views, genetic engineering is regarded as interference with the integrity of individuals and species—that which makes them what they are and is valuable. (For a further discussion of this point, see Dobson 1995.)

There are, then, a number of ethicists who reject genetic engineering on principle—because it is perceived to be unnatural, because it is seen as a reprehensible exhibition of human dominance over nature, or because it interferes with the natures of valuable individuals and species. Such ethical rejection of genetic engineering is absolute; it is independent of the consequences (whether good or bad) that may result from the practice of genetic engineering. Other ethicists do not object in principle but suggest that genetic engineering raises a number of ethical questions—specifically, for our purposes here, environmental ethical questions—in practice. Leaving aside issues of animal welfare, these issues mostly relate to the environmental effects of releasing genetically modified organisms into the environment.

Most generally, it is argued that the effects of releasing genetically modified organisms into the environment cannot be predicted. It is feared, for instance, that genes from genetically modified crops may pass into wild species, causing unpredictable, potentially ecologically damaging consequences. There is also concern about loss of biodiversity, for a number of reasons. Genetically engineered crop plants may prove to be such strong competitors that they eliminate wild species. Crops may be genetically engineered so that they can grow in wild areas that were previously unsuitable for them, driving out native species. Indigenous crop varieties may be replaced by standard genetically engineered forms or genetic clones with no genetic diversity. Other concerns revolve around the possible increase in use of herbicides if crops are genetically engineered to be resistant to their effects.

But there are also, of course, possible positive environmental outcomes from the use of genetically modified organisms. Genetically modified plants may mean fewer rather than more pesticides are used if plants can be engineered to resist particular kinds of diseases; similarly, they may mean less fertilizer is needed if plants can be engineered to grow in poorer soil. Higher-yielding genetically engineered crops may mean that less land is needed for agriculture and more can be set aside for wildlife. Genetically engineered bacteria may have a multitude of uses in environmental cleanup and decontamination.

Despite these possible positive outcomes, some ethicists reject genetic engineering in an environmental context on the grounds that the environmental costs outweigh the benefits. Other ethicists argue that a case-by-case approach should be taken, weighing the environmental risks of any particular piece of genetic engineering against the benefits and applying a precautionary principle where appropriate.

References: Dobson, Andrew. 1995. "Biocentrism and Genetic Engineering." *Environmental Values* 4: 227–239.

Lee, Martin. 1995. "EPA FY1996 Appropriations: Analysis of House-Passed Riders." Congressional Research Report for Congress, November 3, 95-966 ENR.

Lucassen, Emy. 1996. "The Ethics of Genetic Engineering." *Journal of Applied Philosophy* 13(1): 51–61.

Norman, Richard. 1996. "Interfering with Nature." *Journal of Applied Philosophy* 13(1): 1–11.

Walgate, Robert. 1990. *Miracle or Menace: Biotechnology and the Third World*. London: A Panos Dossier.

Wheale, Peter, and Ruth McNally, eds. 1990. *The Bio-Revolution: Cornucopia or Pandora's Box*. London: Pluto Press.

Hunting and Fishing

Hunting and fishing may be defined as the practice of pursuing and attempting to catch other organisms as prey. These activities may be undertaken for a variety of reasons. For instance, hunting may be undertaken for subsistence purposes—that is, to allow the human individual concerned to subsist, to carry on living. It may be undertaken as sport, where the hunted animal is not consumed or otherwise used, or where the consumption is not the reason for the hunting (such as fox hunting in Britain). It may be part of a commercial enterprise. It may also be undertaken for ecological reasons—to control populations of particular kinds of animals; in this instance, it is more commonly called culling. Similarly, there are different kinds of fishing: catch-and-release fishing, for instance, where fish are taken as sport and usually thrown back into the water, and commercial fishing, which itself can take many forms. When considering the ethical questions raised by hunting and fishing, it is important to be clear about the kind of activity at issue.

Hunting and fishing have always been important human activities, primarily for subsistence and commerce but also for sport. Internationally, fisheries produce about 16 percent of total animal protein available in the world

(FAO 1991). However, catches by both developed and developing countries have increased substantially in recent decades, which has led to problems of overfishing in some areas of the Northern Hemisphere and the scarcity of certain species of fish, such as cod and herring. Fishing techniques, too, have a variety of environmental impacts, from the devastating effects of nylon drift nets (some 100 km long) to the loss and disposal of more than 150,000 tons of fishing gear in the oceans each year (GESAMP 1990).

Hunting has generally assumed less importance as a means of basic subsistence owing to the domestication of animals (indeed, the increase in aquaculture may lead to a similar decrease in dependence on wild fishing). For some cultures, subsistence hunting is still of central importance. In the United States, however, hunting is primarily pursued as a recreational activity and focuses on a variety of bird species, especially ducks, geese, woodcocks, snipes, and doves. The Fish and Wildlife Service, which oversees hunting and fishing in the United States, attempts to ensure that the number of animals and birds hunted remains within sustainable limits by carrying out surveys of populations and number of animals hunted.

Ethical Issues Raised   Most hunting and fishing, with the exception of catch-and-release fishing, involves killing, and the ethical acceptability of killing is one of the major questions raised by hunting and fishing. How this killing is regarded depends on the value placed on the life of the individual killed and the context in which the killing takes place. After all, it is by no means generally accepted that the killing, even of human beings, is always wrong: for instance, killing on the battlefield is widely considered to be ethically acceptable. Thus, it is important to ask, Why is killing wrong? Or in this context, Is it wrong to kill when hunting and fishing?

There is a wide range of possible answers to this question. Many of them suggest that hunting and fishing are ethically wrong. Most commonly, such arguments focus not so much on the ethical unacceptability of taking life but rather on the ethical unacceptability of causing suffering. Someone taking this position might argue that animals, in particular mammals, as organisms with complex nervous systems, are capable of suffering intense pain and distress. Hunting, they might argue, causes pain and distress; therefore hunting is wrong.

However, there are problems with some aspects of this argument. For instance, there are occasions when causing suffering might not be wrong, such as when a surgeon performs an operation. But even if we accept the general principle that suffering is wrong, it applies unevenly to different kinds of hunting and fishing. Different species may have varying degrees of sensitivity to pain. Some scientists have argued that fish feel no pain at all; if this were the case, then suffering would not be grounds for thinking that fishing was wrong. Different kinds of hunting may also cause different degrees of suffering. Hunting with hounds as practiced in Britain, with a long chase and a potentially drawn-out death, would certainly generate substantially more suffering than shooting with an accurate shot. But an inaccurate shot could generate more long-term suffering than a hunt with hounds. It is undoubtedly true, however, that most forms of hunting, and perhaps fishing, do have the potential to cause significant suffering. For ethicists such as Peter Singer, who consider the reduction of suffering to be of central significance, this is the prime argument against hunting and fishing.

For other ethicists, however, the wrong in hunting and fishing lies not with the suffering that can be inflicted but with the taking of life itself. Those who argue that animals have rights, for instance, think that killing animals is always wrong, whether or not suffering is involved. Tom Regan (1984), for instance, argues that animals are "subjects of a life": they can have experiences, feel pain, plan their own behavior. Thus, animals have rights, according to Regan, and should not be treated simply as human resources. Hunting an animal violates its rights and is ethically unacceptable.

Not all ethicists argue that killing when hunting or fishing is wrong. A variety of arguments affirm the ethical *acceptability* of hunting and fishing—in certain contexts at least. One of the most significant of these is the environmental or ecological argument that in some circumstances, hunting (or culling) is essential for conservation of ecosystems and species. Culling is normally done when one species has multiplied greatly, perhaps due to the extinction of natural predators, and is threatening to destroy the ecosystem it inhabits. Failure to cull might result in the loss of other species, destruction of habitat, soil erosion, and the eventual starvation of even the destructive species itself. In such circumstances, it can be argued that there is an ethical imperative to hunt. It might even be the case that, in the long run, less animal suffering would result from hunting.

It should be noted that ecological arguments like the above do not support hunting in all circumstances. Where a species is endangered, particularly if it is vital to its ecosystem, hunting can have a detrimental effect on the conservation of ecosystems and species. Drift-net fishing raises further ethical questions here, since the process of fishing for the desired species (which may not be scarce) involves entrapping individuals of endangered species including marine mammals that may also be highly sensitive to pain. In such a case, ecological ethical arguments would oppose any hunting or fishing. It is important, then, to point out that ethical arguments based on ecology, either for or against hunting, are very much context-dependent.

So far, we have considered ethical arguments for and against hunting and fishing based on ecological effects and the effects on the hunted organisms themselves. From these perspectives alone, the majority of arguments oppose the activities of hunting and fishing. However, we have not yet considered the beneficial effects of hunting and fishing on human beings, both humans alive now and those yet to be born. Accepting the fact that there are benefits does not necessarily mean that the arguments against hunting should be discounted but reveals that the issue is a difficult and complex one.

Clearly, hunting and fishing can provide significant benefits for human beings. Food, materials, sport, and wealth are among the most important.

For some peoples, hunting and fishing is essential for survival or for continuance of their traditional cultures. About 60 percent of the developing world derives 40 percent or more of its total annual protein from fish (UNEP 1993). For other people, hunting and fishing may not be essential to their society but may provide income and employment; for some, these activities provide an increase in corporate profits; for yet others, leisure and enjoyment. Can any of these benefits be used to support ethical arguments for or against hunting and fishing?

Ethical arguments for hunting or fishing are strongest where these activities are vital for human survival. Among peoples such as the Inuit, other sources of food and clothing may be difficult or impossible to obtain. In such circumstances, to outlaw hunting would, at the very least, result in the destruction of a culture, possibly even in the loss of individual lives. This situation provides an ethical dilemma even for those who uphold animal rights, since the lives of human rights bearers may also be at stake. The ethical choice one makes in these circumstances would depend on whether one believed that, in a situation of conflict, human rights to live have priority over those of animals.

At the other end of the scale, ethical arguments can be put forward to support hunting and fishing for leisure and pleasure. For those who do not accept that animal suffering or lives are ethically significant, the fact that intense human pleasures can be gained from hunting and fishing is sufficient justification for going ahead. Even those who do consider animal suffering or animal lives to be of some ethical significance can argue that human pleasure outweighs animal suffering. Furthermore, in some countries, such as the United States and the United Kingdom, where wild areas are becoming scarce, hunting and fishing lobby groups can be the strongest campaigners for the protection of wild areas. Thus, provided that the hunting and fishing were not itself damaging to the natural environment, hunting and fishing within sustainable limits could increase environmental protection. However, this argument is not really focused on the debate over hunting and fishing but uses hunting and fishing as a tool to achieve another end: protection of wild areas.

It is already clear that the ethical debate over hunting and fishing is a complex one, but one further factor must be introduced. We have considered the effects of hunting and fishing on animals and ecosystems along with the benefits human beings can gain from these activities. However, we have not considered the question of which human beings should benefit from these activities—in particular, whether future generations of people are entitled to the same benefits (*see* Sustainable Development and Future Generations). This question is particularly acute given current concerns about overfishing. Intensive fishing now may mean that future generations may not be able to enjoy the variety of seafood that is currently available; it may even lead to famine and starvation in areas dependent on the sea for supplies. Thus, human hunting and fishing activities today can have direct implications, not only for the individual animals and humans involved but also for generations of individuals not yet born. This ethical concern has led to the concept of sustainable yields, which means that hunting and fishing are controlled in an attempt to achieve a constant population of the species concerned. This policy aims to ensure that fishing and hunting on the same scale could continue indefinitely. The success of this method depends, of course, on the accurate measurement and prediction of population levels, a difficult and controversial process.

Hunting and fishing, then, raise a number of different ethical issues: of animal suffering and animal lives, of the conservation of ecosystems, of human survival and well-being, of the needs of future generations. Though important ethical principles are involved, it is difficult to come to general conclusions about the rights and wrongs of hunting and fishing.

References: Food and Agricultural Organization. 1991. *Environment and Sustainability in Fisheries.* Rome: FAO Document COFI/91/3.

Regan, Tom. 1984. *The Case for Animal Rights.* London: Routledge.

Singer, Peter. 1973. *Animal Liberation.* (1983 ed.). Wellingborough, England: Thorsons Press.

United Nations Environmental Programme. 1993. *The World Environment 1972–1992.* London: UNEP/Chapman and Hall.

United Nations Group of Experts on the Scientific Aspects of Marine Pollution. 1990. *The State of the Marine Environment.* Nairobi: UNEP.

Natural Resource Depletion

Natural resources can be defined as anything that human beings use or may potentially use that is not of human origin—for instance, oil, coal, metal ores, oceans, and forests (Water Resources and Energy Resources are discussed in separate entries). The depletion of such resources refers to their being used up by human beings.

All living organisms rely on some part of the external world to provide them with the means to carry on living—to breathe, to take in nutrition, and so on. All organisms are, in this sense, dependent on resources outside themselves, and human beings are no exception to this rule. We use a wide range of external natural resources to maintain life and to preserve our societies. Such natural resources can, importantly, be categorized in several ways. One of the key distinctions is that between renewable and nonrenewable resources.

Renewable resources are resources that can be replaced or renewed as they are used. Some renewable resources are inexhaustible (for instance, solar power from the sun). Others can be exhausted if the rate of depletion is greater than that of renewal (for instance, forests are renewable but can be exhausted if the rate of depletion exceeds that of replanting). Still other renewable resources are renewable only to a point—for instance, any particular species is technically renewable in that its members can keep breeding, but populations can fall below a minimum level of viability, leading to extinction and the permanent loss of the resource. In this sense, even some renewable resources can be depleted or exhausted. The expression *sustainable use* is sometimes applied to the use of renewable resources at a level that they can replenish themselves (Daly 1990).

Nonrenewable resources are those that cannot be renewed or replaced; once used, they are gone forever. It is nonrenewable natural resources that are usually being referred to in the context of resource depletion. Daly (1990), however, proposes that talking about the sustainable use of nonrenewable resources can make sense—if they are used up at no greater a rate than their *function* can be replaced by renewable resources. (For example, using up oil is sustainable if substitutes for oil are being developed and introduced at the same rate that the oil is being used up.) This idea of substitution is central to the ethical debate about natural resources, as the next section will indicate.

The depletion of nonrenewable natural resources has been a key issue in environmental debate for at least 25 years. It was one of the main areas of concern, alongside industrialization, population growth, malnutrition, and a generally deteriorating environment, that was highlighted by the 1972 report of the Club of Rome, *The Limits to Growth*. This report, which used a detailed computer model of the world's natural resources, estimated that exhaustion of some of the world's mineral supplies would occur within 50 years, and that most of the remaining supplies would be gone within 150 years (Meadows et al. 1972). The report also expressed concern about lack of freshwater supplies and arable land to feed growing human populations.

After its publication, this report was widely criticized for using inadequate computer models and for failing to take into account the increasing efficiency of resource use and the discovery of new reserves of nonrenewable resources such as oil. However, more recent reports have raised similar questions about global resource depletion. In 1992, the same research team published a follow-up report, *Beyond the Limits*, which restated the argument that human beings were depleting their resources, both by not allowing renewable resources time to replenish themselves and by exhausting nonrenewable resources without developing renewable substitutes. Similar arguments can be found in a range of other recent reports on the use of natural resources, including the 1996 *State of the World* report, produced by the Worldwatch Institute in the United States. Depletion of both renewable and nonrenewable resources is widely regarded as one of the gravest environmental problems facing humanity.

Relevant Ethical Issues. In order to understand the issues raised for environmental ethics by natural resource depletion, it is important to look more closely at economic ideas about what natural resources are. Economists generally describe the world in terms of natural resources that perform particular useful functions for human beings—what is important is what the natural world can do for humans, the benefits it provides. If the natural world is understood in this way, then so long as humans get the benefits, the particular source of the benefits is irrelevant. This is why the idea of *substitution* is so important economically. One resource can be substituted for another without loss provided that the same benefit to humans exists. For instance, humans need paper from wood. Wood can be provided from an old-growth forest or from a new timber plantation. As long as the paper is the same, one resource can be substituted for another without any loss. Human beings need electricity. Electricity can be produced by burning coal in power stations or by wind farms. As long as the same amount of electricity can be produced, one can be substituted for the other without any loss.

Of course, no economist would argue that all natural resources are similarly substitutable. For instance, there are no substitutes for stratospheric ozone, which protects life on earth from the sun's ultraviolet light (Pearce et al. 1989). Neither are there alternatives to fresh, clean water for drinking or for good agricultural land. Economists agree that there are some resources for which humans currently have no substitutes—and the depletion of these is a matter of particular concern.

Furthermore, natural resources may have several functions, some of which are conflicting and some of which are nondestructive. For instance, an old-growth forest may be used for logging, but it may also provide a place where human beings can go to seek solitude and beauty. The planting of a forestry plantation may succeed, in logging terms, in substituting for the old-growth forest, but it is unlikely to provide the same esthetic pleasures to human beings. Thus, the loss of the forest may only be substitutable in terms of some functions, and not others.

However, many environmental ethicists go much further than this. They argue that understanding the earth merely as a resource is inadequate, even where the resource is regarded as providing the nondestructive pleasures of beauty or solitude to human beings. They argue that the earth (or some elements of it) has value as well, independent of its usefulness to human beings, a value sometimes called intrinsic value (discussed further in the introduction to this book). This value cannot be replaced by artificial substitutes. The destruction of an old-growth forest, such ethicists argue, is a loss of value even if it were replaced by another wood-providing plantation, and even if, in addition, humans were to be provided with satisfactory substitute experiences by virtual reality machines that re-created the sensations of being in old-growth forest. Elements of the natural environment are not, from this perspective, substitutable. The depletion of natural resources can be regarded not only as a loss to present human beings and to future generations of human beings who will not have access to such resources, but also as a loss of value in the world.

Ethicists, however, differ over exactly how this value is understood and what it is associated with. Those primarily concerned with the value and well-being of nonhuman animals, for instance, argue only that *animals* should not be regarded as resources. The felling of an old-growth forest might then be ethically problematic for two reasons: first, by killing the animals that lived there, we are assuming that animals are substitutable resources; second, in destroying the forest for their own benefit, humans would be unjustly denying access to the forest to animals whose resource it is also (indeed, some ethicists would argue that it is as much the resource of animals as human beings). Other ethicists—those adopting a deep ecology perspective, for instance—would argue that this approach is really only an extended resource-based understanding of the natural world. They might argue that all individual living organisms, or all wild ecosystems and species, are irreplaceably valuable and should not be treated merely as resources, whether for human beings or for human beings and animals as well (Devall and Sessions 1985).

Of course, even those adopting such ethical approaches do not argue that minerals such as iron, oil, or gold have value in themselves (except as they may be integrated into valuable ecosystems). However, the extraction and the use of such minerals may have substantial negative effects upon the environment. Resource depletion may intensify this environmental damage as stocks of mineral resources become harder to find, are located in more remote places (such as Antarctica), or become more difficult to extract. In this way resource depletion may be of further concern to environmental ethics.

In summary, then, the idea that the earth may be primarily thought of in terms of natural resources is questioned by many environmental ethicists, who argue that it is precisely this underlying philosophical view of the world that has caused many of the environmental problems that human beings currently face.

References: Daly, Herman. 1990 "Towards Some Operational Principles of Sustainable Development." *Ecological Economics* 1: 1–6.

Devall, W., and G. Sessions. 1985. *Deep Ecology*. Layton, UT: Peregrine Books.

Meadows, Donella H., et al. 1972. *The Limits to Growth*. London: Pan.

———. 1992. *Beyond the Limits*. London: Earthscan.

Pearce, David, et al. 1989. *Blueprint for a Green Economy*. London: Earthscan.

Worldwatch Institute. 1997. *State of the World Report, 1996*. London: Earthscan.

Nuclear Power and Weapons

Nuclear power is power or energy produced by a controlled nuclear reaction involving the fission or fusion of atomic nuclei. Nuclear weapons are weapons intended to cause mass destruction based on a nuclear reaction.

Einstein's famous formulation for the conversion of mass into energy led, over several decades, to scientific acceptance of the possibility of splitting the nucleus of the atom. In 1939, the U.S. government began work to achieve this end with a project (eventually called the Manhattan Project) to build an atomic weapon for use in World War II (Schell 1982). The atomic bombs resulting from this project were dropped, to hugely destructive effect, on the Japanese cities of Hiroshima and Nagasaki in 1945.

Since 1945, nuclear programs have been pursued by about 26 countries. The purpose of the majority of these programs has been generating energy from nuclear power (although in some countries, such as the United Kingdom, the civil nuclear program was, for some years at least, inextricably linked with the military program). By 1989, 426 commercial nuclear power stations were in operation globally, generating about 17 percent of the world's electricity (Roberts 1993). A smaller number of countries have also pursued nuclear weapons programs, creating large numbers of increasingly destructive nuclear weapons with the potential to cause mass destruction and death. By the early 1990s, more than 50,000 nuclear warheads had been deployed or stockpiled worldwide (Renner 1994).

Both civil and military uses of nuclear power can have harmful environmental effects. Nuclear power stations may deliberately discharge or accidentally release low levels of radioactive material to the environment, the cumulative effects of which on human health and on the environment remain uncertain (see also the entry on Waste). They also create the risk of devastating nuclear accidents such as the one that occurred at the Ukrainian plant Chernobyl in 1986, that killed 31 people outright and contaminated huge areas of land with radioactive cesium.

Clearly, however, the most destructive potential for nuclear technology globally lies in the possibility of a large-scale nuclear war, though scientists dispute the environmental consequences of such a war. Some scientists have predicted a "nuclear winter," where dust particles thrown up by explosions would prevent heat and light from the sun from reaching the earth (see Schell 1982). Others have disputed this claim. But whatever the likelihood of a nuclear winter, the large-scale use of nuclear weapons has the potential to kill more than 1 billion people and perhaps to eliminate the human species (omnicide), as well as causing the extinction of many nonhuman species (see Ehrlich et al. 1984, Harwell 1984). Any large-scale nuclear exchange would indisputably have devastating environmental consequences. The likelihood of such an exchange seems to have receded with the breakup of the former Soviet Union and the signing of a variety of nuclear disarmament agreements. But the possibility of a nuclear exchange remains alive as long as the weapons and the ability and willingness to produce them remain.

International Agreements The possession of nuclear technology is highly regulated, both internationally and nationally. The International Atomic Energy Authority was established in 1956; its stated purpose is to "accelerate and enlarge the contribution of atomic energy to peace, health and prosperity throughout the world" (IAEA 1973). This body is the key regulator for atomic power internationally.

Several important international agreements on nuclear power and weapons have been signed since the establishment of the International Atomic Energy Authority. Central among these is the 1958 Euratom Agreement, drafted by European countries in order to "create conditions necessary for the establishment and growth of nuclear industries" (Behrens and Donnelly 1996). Fifteen European countries signed the agreement, eight of which have functioning commercial nuclear power plants. All signatories have also signed the 1978 Nuclear Non-Proliferation Treaty and are members of the International Atomic Energy Authority. The United States joined the agreement in order to promote sales of U.S. nuclear technology in Europe and renegotiated continued membership in 1995.

Relevant Ethical Issues The generation of nuclear power and the production, storage, testing, and possible use of nuclear weapons raise some of the most fundamental ethical questions facing human beings today. In some areas the questions raised by nuclear power generation are very similar to those raised by nuclear weapons production, but the two activities also create their own distinct ethical problems.

As Paul Thompson (1984) points out, the ethical debate surrounding the generation of nuclear power concerns two key areas: *need* and *safety*. These questions are, in part, factual (requiring, for instance, information on energy use and alternative methods of energy production, though even this factual information may be contested). Even so, weighing these facts and assigning value to them is clearly an ethical matter. At one extreme, some radical environmentalists—many deep ecologists, for instance—reject the very idea of nuclear power, for a variety of reasons. Most fundamentally, these radical environmentalists reject the idea of materialistic, affluent human societies with such high energy consumption that they need to be underpinned by nuclear power. Secondly, they might argue that the production of nuclear power is large rather than small scale, and that it centralizes energy generation and necessarily entrusts control to an elite group of scientists, bureaucrats, and security police (in contrast to simple, small-scale technology democratically controlled by local people). Thirdly, they would argue, it poses serious risks of irreversible harm to the life and health of human beings (both present and future), to animals and other organisms, and to ecosystems and the environment in general. These environmentalists reject the use of nuclear power in principle and in all circumstances.

Other ethicists refuse to reject nuclear power *in principle*, acknowledging that substantial human benefits may be gained from the use of nuclear power, especially in countries that have few other ways of generating energy or are historically dependent on nuclear power. However, those who accept such positions do not necessarily agree with the generation of nuclear power in practice, in the circumstances in which we now find ourselves. They may argue that the risk of harm to present and future human beings, animals, and the environment from an accident at a nuclear power station (or from long-term leaks or disposal of waste) outweighs the benefits gained from the generation of the power. Relevant to this kind of ethical calculation are the environmental costs of nuclear power as opposed to the environmental costs of other forms of electricity generation—given, for instance, the carbon dioxide emissions produced by coal- and oil-fired power stations (*see* Climate Change).

Nuclear weapons raise an even greater range of ethical issues. Clearly, the *use* of nuclear weapons must raise ultimate questions for environmental ethics, since a nuclear war might eliminate all or much of life on earth. Although we are not certain exactly how serious its effects would be, it is likely that nuclear war would kill millions of individual organisms, eliminate species—including possibly the human species—and damage or destroy all ecosystems. As Michael Fox (1987) maintains, nuclear war would be the ultimate environmental crisis. If this is not a concern for environmental ethics, then what is?

From almost all perspectives in environmental ethics, the instigation of a nuclear war must be an act of supreme ethical wrong. For example, from a rights perspective, the rights to life of millions of living beings would be removed. From a utilitarian perspective, the happiness in the world (both human and nonhuman) would be drastically reduced and replaced by suffering, assuming any organisms remained alive at all. An extreme misanthropist might argue that a nuclear war that eliminated all humans but left some other living organisms to reevolve into a living world without humans might be a good thing, although (for a variety of reasons) such a position would be very hard to maintain!

The potentially devastating consequences of nuclear war have led to ethical opposition to the existence of nuclear weapons from a variety of environmental, feminist, and peace groups. Ecofeminists in particular have opposed the construction and stockpiling of nuclear weapons: "Nuclear madness needs to be taken seriously as a madness, that is as a craziness which has delusion, denial and disassociation at its core" (Warren 1994). Others have argued that as long as nuclear weapons are never used but kept only as deterrents, the consequences of constructing and storing them are preferable to not having them (or not having them when someone else has them). However, nuclear weapons cause environmental (and human) harm even when they are not deliberately used, in terms of uranium mining, nuclear materials processing, nuclear waste storage, weapons testing, and the continuing risk of accidents. Even nuclear disarmament agreements cause problems, since the disposal of the nuclear material from weapons is as problematic as disposal of other forms of nuclear waste. Given that their purpose is to cause mass destruction, and that they cause environmental damage even when not used, it is difficult to make a case from any perspective in environmental ethics for the retention of nuclear weapons.

References: Behrens, Carl, and Warren Donnelly. 1996. "Euratom and the United States: Renewing the Agreement for Nuclear Co-operation." Congressional Research Service Report for Congress, 26 April, IB96001.

Erlich, Paul, et al. 1984. *The Cold and the Dark: The World after Nuclear War*. New York: W. W. Norton.

Fox, Michael. 1987. "Nuclear Weapons and the Ultimate Environmental Crisis." *Environmental Ethics* 9(2): 161–179.

Harwell, Mark. 1984. *Nuclear Winter: The Human and Environmental Consequences of Nuclear War*. New York: Springer-Verlag.

Henderson, Hazel. 1981. "The Challenge of Decision Making in the Solar Age." In David Brunner, ed., *Corporations and the Environment: How Should Decisions Be Made?* Los Altos, CA: William Kaufmann.

International Atomic Energy Authority. 1973. Article 2 of statute (amended 1973).

Renner, Michael. 1994. "Cleaning Up after the Arms Race." In Worldwatch Institute, *The State of the World 1994*. London: Earthscan.

Roberts, L. E. J. 1993. "Case Study: Nuclear Power." In Samuel Berry, ed., *Environmental Dilemmas: Ethics and Decisions*. London: Chapman and Hall.

Schell, Jonathan. 1982. *The Fate of the Earth*. London: Pan.

Thompson, Paul. 1984. "Need and Safety: The Nuclear Power Debate." *Environmental Ethics* 6(1): 57–71.

Warren, Karen. 1994. "Towards an Ecofeminist Peace Politics." In Karen Warren, ed., *Ecological Feminism*. London: Routledge.

Population

Population is generally defined as the number of human beings alive in the world at any particular time.

Although global human population has steadily increased for millennia, it has risen sharply since the 1950s. Currently, the total annual increase in world human population is 1.7 percent, that is, about 90 million people per year (Warner et al. 1996). The population of developed countries is expanding at about 0.6 percent annually, while that of developing countries is increasing at about 1.9 percent. Predictions of what might happen to global population in the next 100 years vary. Current global population is estimated to be approximately 5.7 billion. In 1992, the World Bank published a report on population with three possible trajectories for population increase. The most likely model predicted population stabilizing at about 12.5 billion in 2050; the highest trajectory allowed for the possibility of population increasing to 22 billion in the next 150 years (World Bank 1992).

Although the highest levels of population growth are occurring in developing countries, substantial increases are taking place in developed countries as well. For example, total U.S. population was estimated to be 262 million as of March 1995, an increase of 5.3 percent from 248.7 million in 1990, with average population growth of just under 1.1 percent a year since 1990. Estimates suggest that by 2050, the U.S. population could be 392 million—nearly a 50 percent increase from 1995 (Williams 1995). This increase is due to the combination of relatively high birthrates and levels of immigration. However, this population growth has not been and will not be evenly spread over the United States. States in the West and South have shown higher rates of growth.

International Population Policy International political concern about expanding human population was first voiced during the 1970s. In 1974, the United Nations passed General Assembly Resolution 3345 (XXIX), which called for further studies on the relationship between population and resources. The first International Conference on Population was held in Mexico City in 1984. Delegates at this conference recognized the complex relationships between population, resources, environment, and development, noting a disequilibrium between the rate of change of population and the rate of change in available resources. The conference acknowledged that rapid population growth could aggravate environmental and resource problems such as soil erosion and desertification (U.N. 1984). A second International

Conference on Population and Development was held ten years later in Cairo. At this conference, delegates were asked to sign a Program of Action on population, which outlined ways of reducing population growth: increasing economic development in the poorer nations, providing women with access to family planning, and improving women's education and status. The promotion of economic development in poorer nations received virtually unanimous support, but some religious groups (such as representatives of the Roman Catholic Church) opposed increased availability of family planning. Others, including some Muslim organizations, opposed changing women's status; this issue led to the withdrawal of some Islamic countries from the conference. Although representatives from many countries signed the Program of Action, the program is not legally binding.

Key Debates over Population   The possible consequences of human population growth have been hotly contested since the writer Malthus published his *Essay on Population* in 1798. In this book, Malthus argued that "the power of population is indefinitely greater than the power of the earth to provide subsistence for man" (1927 ed.). This view—that inexorable human population increase must outstrip the resources available to provide food, clothing, and shelter and must ultimately result in famine, disease, and war—has been developed and adapted by a number of modern writers. These writers draw on the ecological concept of carrying capacity—the number of individuals of any species, human or nonhuman, that a given environment can support or "carry" without depleting its base—and argue that the earth's carrying capacity for human beings is limited. Continued population increase will inevitably result in hunger, malnutrition, disease, and war. The most prominent exponents of such views have been Garrett Hardin and Paul Ehrlich. Hardin's 1968 paper, "The Tragedy of the Commons," has been particularly influential in this debate (as well as in debates over the use of natural resources and the production of pollution). Hardin argued that, as rational beings, every individual human acts selfishly to better his or her own position—in this instance, to pursue his or her freedom to reproduce. But the effect of everyone doing this is an excessive human population that cannot be supported by available resources. Disaster will inevitably follow.

In contrast to these neo-Malthusian views are the arguments of commentators who argue that technological innovation (for instance, new methods of agriculture, use of biotechnology) means that there is no "natural" limit to human population. If resources become scarce, then the price of such resources will be forced up, and the rise in price will encourage the development of substitutes, which will then enter the marketplace. These population optimists do not regard the rise in human population as potentially disastrous but rather as providing a challenge to human technological ingenuity.

Recent international political debate has recognized that population increase may pose difficulties, especially in particular places and to particular groups of people. However, little progress has been made toward dealing with the issue, since so many delicate cultural questions such as religious practice, human rights, gender issues, and distribution of wealth and consumption internationally are raised by the population debate.

Relevant Ethical Issues Obviously, a variety of human ethical issues are raised by the population debate. For those tending toward a Malthusian position, two central issues emerge: whether humans have, or should have, the right or the freedom to reproduce, and whether aiding those already suffering from hunger or malnutrition will lead to an increase in population that will, in turn, increase suffering (see, for instance, Hardin 1977). Others have questioned the argument that the ethical problem concerns freedom to *reproduce* (especially in developing countries); instead, they suggest, the problem concerns the freedom to *consume* (particularly in the richer, developed nations). The population optimists do not see population increase as generating any special human ethical problems at all.

However, human population increase has been seen as problematic by a number of *environmental* ethicists. The focus of their concern has been the expansion of human populations into previously wild areas and the accompanying increase in world biomass constituted by domesticated rather than wild species of animals and plants. Westing (1981) points out that in 1980, 20 percent of world biomass was constituted by human beings and domesticated animals and plants. He estimates that in order to feed growing human populations and meet higher standards of living by 2010, 40 percent of global biomass will be composed of human beings and domesticated plants and animals, rising to 60 percent in 2050. Such an increase in human presence and activity on the earth will inevitably mean the loss of many wild individuals, species, and ecosystems along with areas we currently think of as wildernesses. The price for high human population is therefore the loss of wild areas (Warner et al. 1996). Clearly, this situation must be of concern to the environmental ethicists who focus primarily on the value and protection of wilderness and wild nature. It is also of concern to those ethicists who argue that organisms other than human beings have rights and "might also have some claim of a right of access to the means of their life" (Hayward 1995). Deep ecologists in particular have argued the human population must be reduced in order to make space for the flourishing of wild nature. One of the main ethical principles accepted by many deep ecologists affirms just this: "The flourishing of human life and cultures is compatible with a substantial decrease in the human population. The flourishing of nonhuman life requires such a decrease" (Devall and Sessions 1985).

This position, however, is controversial, even among environmental ethicists. Some have argued that it is misanthropic and avoids difficult political issues of social equity and international distribution of wealth. Others have argued that it reveals the predominant bias in environmental ethics toward *wild* environments when agricultural landscapes and urban townscapes are also environments that can be understood as valuable, albeit in different ways. Thus, human population growth is a subject of considerable debate inside as well as outside environmental ethics, and owing to the different ethical premises of the contributors to the debate, it seems unlikely that any consensus will emerge in the near future.

References: Devall, William, and George Sessions. 1985. *Deep Ecology: Living as If Nature Mattered.* Salt Lake City, UT: Peregrine Smith.

Ehrlich, Paul. 1968. *The Population Bomb.* New York: Ballantine.

Hardin, Garrett. 1968. "The Tragedy of the Commons." *Science* 162 (13 December): 1243–1248.

———. 1977. "Lifeboat Ethics: The Case against Helping the Poor." In W. Aiken and H. La Follette, *World Hunger and Moral Obligation.* Englewood Cliffs, NJ: Prentice-Hall.

Hayward, Tim. 1995. *Ecological Thought.* London: Routledge.

United Nations. 1984. Report of the International Conference on Population. New York: UN.

Warner, Stanley, et al. 1996. "Global Population Growth and the Demise of Nature." *Environmental Values* 5(4): 285–303.

Westing, A. 1981. "A World in Balance." *Environmental Conservation* 8(3): 177–183.

Williams, Jennifer. 1995. "The U.S. Population: A Factsheet." Congressional Research Service Report to Congress, 12 June, 95-705 GOV.

World Bank. 1992. *World Development Report.* New York: Oxford University Press.

Sustainable Development and Future Generations

Sustainability is usually defined as the ability to keep going indefinitely, or at least for an extensive period of time. The definition of development is more difficult; it is sometimes defined as bringing out what is latent, working out potentialities. What exactly might be meant by the expression *sustainable development* in a political context is much disputed (for example, a distinction between "strong" and "weak" interpretations is often made). The most commonly used definition of sustainable development (and the one that will be adopted here, though not without comment) is that proposed by the World Commission on Environment and Development (WCED) in its 1987 report, *Our Common Future:* "Sustainable development is development that meets the needs of the present without compromising the ability of future generations to meet their own needs." The definition of a generation is somewhat vague, but it normally refers to people living at the same age or period, and successive generations are usually defined as being about 30 years apart. Future generations are those generations of people who do not exist at the same time as present generations; that is, they have not yet been born.

The concept of sustainability and the associated idea of sustainable development have become increasingly important in international political discussion. It was featured in the publication of the World Conservation Strategy in 1980; the 1987 World Commission on Environment and Development report, *Our Common Future;* and the proceedings and agreements of the 1992 Earth Summit held at Rio de Janeiro, Brazil. As can be seen from the definition above, discussions about sustainability have tended to focus on two issues: first, on meeting the needs of people who are presently alive, and second, on not undermining the ability of future generations of people to achieve acceptable standards of living themselves. Discussions about sustainable development in this political context address four key issues: *population*, *consumption*, *resource use*, and *pollution*. (For further discussion of these topics, see separate entries on Population, Natural Resource Depletion, Water Resources, Waste, and Atmospheric Pollution.) A sharp upward movement in any one of these four factors, it is argued, might threaten the well-being of presently existing human individuals and might also undermine the sustainability of human society in the long term, thus affecting the ability of future generations to live well—or to live at all.

However, because we cannot know the future, there is clearly a good deal of uncertainty concerning the long-term sustainability of particular policies. Population optimists, for instance (*see* Population), argue that even quite large increases in population may be sustainable, given the possibility of technological innovation in agriculture. Others argue that substitutes for depleted resources will be discovered or created, and high resource use will become sustainable. In short, there is little agreement about what kind of policy and what sort of development is, in fact, sustainable.

Relevant Ethical Issues Sustainable development, as defined in the WCED document, has two main ethical thrusts. One is toward social justice, the other toward future generations. The idea of sustainable development implies that humans are ethically obligated both to assist the presently needy and to consider the well-being of future generations. Interestingly, a distinction can be made between what is owed ethically to the presently needy and to future generations. For those who are needy in the present, the definition suggests an obligation to meet their needs. For those in the future, the ethical obligation is to avoid inhibiting their ability to meet their own needs. Thus, sustainable development understood in this way entails a positive obligation to *assist* present generations but a negative obligation *not to hinder* future generations (by leaving them a legacy of toxic soil, for example, so that they are unable to grow crops). But among ethicists, neither of these two obligations is universally acknowledged; where ethicists do acknowledge such obligations, a variety of different reasons are given for doing so. (For those interested in pursuing this debate, see the entry on Population and also Palmer 1994, de-Shalit 1995.)

What is interesting for environmental ethics is the relationship between the concept of sustainable development and the environment. Sustainable development has moved to the center of the debate about human beings and the environment—emerging, for instance, as the key theme from the 1992 Rio Earth Summit. Yet, curiously, in the popular definition of sustainable development we have been using, the environment is not directly mentioned at all. Indirectly, the environment is perceived as important because a healthy natural environment is usually thought to be vital for the well-being of present and future human beings.

Present human beings have the capacity to endanger future human beings in a variety of environmental respects: by depleting resources, by storing radioactive waste unsafely, by diminishing biodiversity, by bringing about climate change, and by causing other kinds of pollution. (de-Shalit [1995] explores in more detail these and other ways we may affect future generations.) A land polluted with toxic waste, for instance, might be bad for humans in several ways. They would be unable to produce crops on it, thus inhibiting their ability to provide for their own food needs. Living near the land and breathing dust blown from it might make them sick and even kill them or damage their genetic heritage. The loss of species on the poisoned land might deprive future people of the esthetic pleasure they might derive from the variety of species that the current generation of people enjoys. For these reasons, the environment is ethically important in terms of sustainability; with a damaged environment, it might be impossible to meet the needs of present people or for future people to meet their own needs.

However, many environmental ethicists would find endorsing this interpretation of sustainability difficult. It understands the environment to be purely a resource for meeting human needs rather something that has value in itself. It suggests that, provided means could be found to meet the needs of present people without damaging the prospects for future people, environmental protection would not matter. If food could be grown in special factories, air filters could be used to prevent the inhalation of toxic dust, and as much pleasure could be gained from theme parks and virtual reality rides as from watching wildlife, then the pollution of land with toxic waste would not matter ethically. Human needs would be met.

For environmentalists who think that nonhuman living beings, ecosystems, or species are valuable in ways not related to human use, this ethical approach is unacceptable. It focuses value entirely on human beings. As Holmes Rolston (1991), one of the best-known environmental ethicists, comments, "Let's face it, sustainable development is irredeemably anthropocentric." For this reason, some environmental ethicists—while acknowledging that we have ethical obligations both to current needy people and to future generations—reject the idea of sustainable development altogether. Others have defined sustainable development in an ecological way, to include not only human society but biodiversity, ecological integrity, and ecosystemic processes. This definition affirms the value of sustaining current biodiversity and evolutionary and ecosystemic processes, even where they are not directly useful to present or future human beings.

Such an expansion of the definition of sustainable development does not, however, completely resolve the ethical problems created by use of the expression. It is clearly inevitable, in a world with a growing population where a commitment is acknowledged to assist the needy, that there will be occasions when the sustainability of human society conflicts with the sustainability of ecosystems or ecosystemic processes. If human sustainability always wins in such conflicts, the affirmation of the intrinsic value of ecosystems would be little more than lip service. Yet where great good can be brought to the human needy by damaging an ecosystem, is it ethically acceptable to favor the ecosystem over the needy? This kind of dilemma—addressed in a variety of ways by environmental ethicists (see, for instance, Rolston 1995, Brennan 1996)—lies at the heart of the debate about sustainable development in environmental ethics. How to balance ethical consideration for the human needy, for future people, and for animals and the nonhuman environment is, and will remain, a critical concern for environmental ethicists.

References: Brennan, Andrew. 1996. *Poverty, Environment and Ethics*. Unpublished.

Palmer, Clare. 1994. "Some Problems with Sustainability." *Studies in Christian Ethics* 7(1): 52–62.

Rolston, Holmes. 1991. "The Wilderness Idea Affirmed." *The Environmental Professional* 13: 370–377.

———. 1995. *Conserving Natural Value*. New York: Columbia University Press.

de-Shalit, Avner. 1995. *Why Posterity Matters: Environmental Policies and Future Generations*. London: Routledge.

World Commission on Environment and Development. 1987. *Our Common Future: Report of the World Commission on Environment and Development*. Oxford: Oxford University Press.

Tourism

Tourism is the practice of short-term travel for leisure and pleasure (as opposed to work). One of the fastest-growing industries in the world, tourism is boosted by increasingly easy access to airplanes and motor vehicles. It is currently the second largest contributor to world trade, outstripped only by oil. In 1990, there were 439 million international tourist arrivals, a figure estimated to reach 937 million by the year 2000 (Cater and Lowman 1994). Despite these high figures for international tourism, the majority of tourist journeys are still domestic, and the majority of both domestic and international tourist journeys originate within Europe, North America, and Japan. While the continuing increase in tourism is predictable, fashions in tourist travel are not—although recently the number of long-haul, intercontinental tourist journeys has steadily increased.

It is not surprising, then, that the impact of tourism, both on particular local areas and on global environment and culture, is enormous and increasing. The arrival of large numbers of foreign tourists in any single area can have a variety of social and environmental effects. International tourism can bring foreign currency into countries that need it; the development of tourist facilities may improve conditions for local residents; both tourists and host populations may gain from social mixing; attempts may be made to protect sites of cultural and natural interest for tourists; and tourists may gain a heightened awareness of environmental issues from visiting areas of fragility and beauty. On the other hand, tourism may destroy local cultures by introducing alien ideas and practices; create a service culture of dependency; introduce awareness of economic disparities; create stress and overcrowding both for tourists and for host populations; cause the development of environmentally damaging hotel, leisure, and transportation structures (including roads and golf courses); and contribute to localized pollution. Globally, the tourist industry directly employs many thousands of individuals and indirectly allows for the livelihoods of many more. But it also inevitably entails increased use of motor vehicles and jet engines, which consume fossil fuels and create pollution.

It would be misleading, however, to suggest that all forms of tourism are the same or that they have similar social and environmental impacts. Tourists range from individual backpackers and cyclists to those participating in what is called "mass tourism," characterized by high numbers of visitors to developed resort areas, often as part of package tours. More recently, partly in reaction to mass tourism and its perceived environmentally destructive nature, an increasing number of tourists have opted for nature tourism or ecotourism. Although no accepted definition of ecotourism exists, ecotourist vacations generally emphasize observation and enjoyment of natural environments and encourage (to differing degrees) less environmentally destructive tourist behavior.

Relevant Ethical Issues. Tourism raises a variety of issues for environmental ethics. At one extreme, it might be argued that tourism is ethically unacceptable for those concerned about the natural environment. At the other, it might be argued that some forms of tourism (primarily ecotourism) are positively desirable from an ethical point of view.

One of the main environmental ethical arguments against tourism can be based on the way tourism markets the environment and the way tourists regard it. Some sociologists have argued that tourism turns the environment into a center for human consumption, which may take a variety of forms (Urry 1995). Most often it is visual (what Urry calls "the tourist gaze"), but it may also be literal (tourists can deplete and exhaust a place—for example, tourist hotels may destroy the very coral reefs that attracted tourists in the first place by pumping raw sewage from tourists into the sea). From some perspectives in environmental ethics, this whole approach to the natural world is problematic, even where no physical harm is caused to the environment being visited. Even visual consumption of the environment, it might be argued, is treating the environment as a human resource and valuing it because it gives humans pleasure. It does not recognize that the environment may have value in itself, regardless of whether we can view it or not. For this reason it might be argued that tourism (especially where it entails visiting fragile natural environments) is based on an inadequate, or at least incomplete, understanding of the value of the natural world.

This position, however, is extreme. There are plenty of arguments that run counter to it. Most positively, visiting an environment does not necessarily mean that it is valued *only* as a source of one's own enjoyment. Indeed, tourist enjoyment may be generated precisely *because of* the recognition that such valuable environments exist and will continue to do so, even when the tourist returns home. Tourism may also allow people the opportunity to think about the new environment they are visiting, to contrast it with their home environment, and to make judgments about environmental quality and desirability in ways that would not be possible were they unable to travel.

Certainly, the desire to travel to see an environment (such as, say, a cloud forest or a high mountain plateau), even at the risk of harming it, may indicate that human pleasure is being valued above any intrinsic value in the environment itself. After all, the environment would thrive as well or better if the tourist remained at home—and perhaps gave the travel money to an organization dedicated to its protection. But this does not mean that the tourist does not value the environment at all; it means that the tourist considers the huge personal benefit that will accrue from visiting the place to outweigh the small extra degree of damage that his or her personal visit makes.

This situation raises one of the questions central to ethics—but posed acutely by tourism: *What if everyone does it?* The pattern of tourist activities over the past 50 years indicates that tourist resorts have a cycle of growth. A resort begins with a few travelers (surprisingly, perhaps, the first explorers are the very rich and the very poor), expands to become a fashionable place for mass tourism, then wanes in popularity as new resorts become more fashionable. The exposure to mass tourism and the constant creation of new, fashionable mass resorts causes huge environmental damage (damage enhanced by the fact that most new resorts are farther from the tourist's home and require more transportation than older resorts). As with air pollution (*see* Atmospheric Pollution), this pattern of development raises questions about cumulative harm, where one person's behavior in itself is insufficient to cause harm, but the behavior of individuals en masse causes substantial environmental damage.

A different area of environmental ethical controversy concerns particular kinds of nature tourism where tourists from generally rich countries visit national parks and other kinds of wildlife reserves in generally poorer countries. National governments are disposed to create such parks and reserves to attract rich visitors into the area. But in order to set aside areas as wildlife reserves, governments may forcibly remove native peoples who regard the land as their home. Thus, large swaths of land in less-developed countries may be accessible only to nature tourists, leaving native peoples to suffer dispossession and poverty. Such schemes may have substantial wildlife benefits (especially in the protection of endangered species), but the local people clearly pay a high social cost. Many ethicists now question whether the human cost of such wildlife protection is too high, reflecting a devaluation of human life in developing countries.

In response to the wide range of ethical concerns about the environmental and social impact of tourism, many travel companies have adopted codes of ethics to guide their practice. In 1980, the World Tourist Organization published a declaration at Manila. It maintained that tourism should "protect and respect the natural environment, the values of tourism and the natural, social and human environment." But among critics of tourism, such declarations and ethical codes are received with skepticism. Ethical codes, they argue, are merely tidying up the edges of what is, fundamentally, an environmentally destructive industry.

Tourism is inevitably an ethically controversial area for environmentalists. It has the potential both to inspire environmental protection and cause environmental destruction, to broaden and to narrow the minds of tourists and the opportunities of host populations. It may be regarded as the selfish and self-indulgent luxury of the rich or as a positive, regenerating force for good (issues discussed by Krippendorf 1987). From the perspective of many approaches to environmental ethics, tourism must remain at best an ambiguous phenomenon.

References: Cater, Erlet, and Gwen Lowman. 1994. *Ecotourism: A Sustainable Option?* London: Wiley.
Krippendorf, J. 1987. *The Holiday Makers*. London: Heinneman.
Urry, John. 1995. *Consuming Places*. London: Routledge.
World Tourist Organization. 1980. *Manila Declaration*. Manila: WTO.

Transportation

Transportation services are increasingly important in the lives of individuals internationally. In many developed countries, transportation industries—in particular those centered around the automobile—are linchpins of the economy. In the United States, for example, 5 percent of all workers are employed directly by the automobile industry (Sissine 1996). In developing countries, transportation infrastructure has generally been poorly developed, largely because demand for transportation is closely related to the level of national economic activity (UNEP 1993). However, countries that are expanding economically are in the process of increasing their transportation networks, in particular their road structures.

Road transportation is the biggest consumer of energy of all transportation modes (UNEP 1993). It is also the biggest producer of environmental impacts. In the United States, for instance, automobiles produce 50 percent of the ground level ozone in urban areas and 15 percent of carbon dioxide (a major contributing gas to enhanced global warming). They also account for 37 percent of oil consumption (*see* Atmospheric Pollution and Energy Resources). In practice, this high level of oil consumption adds to the U.S. trade deficit, since $45 billion of oil is imported each year, which equals 30 percent of the U.S. trade deficit (Sissine 1996). Recent government-industry initiatives, such as the 1993 Partnership for a New Generation in Vehicles, have focused on producing energy-efficient vehicles that can achieve 80 miles per gallon. Increased energy efficiency will reduce ground-level ozone and carbon dioxide emissions and protect the national economy against oil price increases from overseas. However, in Europe at least, increases in fuel efficiency have been overwhelmed by an increase in vehicle miles per person per year.

Road construction, too, has huge environmental impacts, especially when roads are built in formerly undeveloped areas. Obviously, the construction requires considerable land, which in itself is often problematic. In the United Kingdom, for instance, to avoid areas of existing human habitation, many proposals for new roads have crossed areas designated by the British government as areas of Outstanding Natural Beauty or Sites of Special Scientific Interest. Alongside this difficulty, the opening of a new road creates enormous problems for existing wildlife. The number of animals killed by motor vehicles increases. Biological communities can be split or broken up and wildlife migration corridors disturbed. Pollution—not only vehicle exhaust but also oil and salt runoff from the road—increases, damaging plant communities near the road and the organisms that depend on them. The area becomes more easily accessible to humans, which may mean more tourists visiting the area, creating a pressure for visitor facilities (food, water, toilets, and therefore sewage disposal) as well as an increase in noise, trampling, and litter. It may also mean—especially in tropical forests—that the area is opened up for agricultural use and for human residency.

Similar environmental problems attend the construction of any nonurban transportation network. They are also a feature of air transportation, which takes vast areas of land for airports and runways, is a significant energy consumer, and creates substantial pollution—and is expanding rapidly. (Between 1972 and 1990, the number of passenger kilometers of air travel more than tripled globally.) Indeed, all major forms of transportation—other than walking, cycling, and using animals—consume natural resources, create pollution, and require infrastructures that involve substantial environmental damage. For this reason, transportation is a central environmental concern.

Relevant Ethical Issues Modern transportation raises a number of profound philosophical—and ultimately ethical—issues about many Western lifestyles. Certainly, modern transportation systems—road, high-speed rail, and air—allow people with sufficient resources to travel much further and faster than would have been possible 50 years ago. However, research suggests that this increasing sophistication in transportation networks has not reduced the total amount of time spent traveling, but rather has increased the distances individuals must travel to perform the same functions (working, shopping, pursuing leisure). U.S. statistics, for instance, show that the average number of vehicle-kilometers per household per year expended on shopping trips increased 88 percent between 1969 and 1990, from 1,495 to 2,804 (USFHA 1992). As transportation expert John Whitelegg (1993) comments, "When people save time, they use it to buy more distance." But, as the writer Ivan Illich pointed out as early as 1974, cars in particular have hidden time costs involved in maintenance, cleaning, searching for parking spaces, and working to pay for, to buy, and to run them.

These changes have important consequences, both socially and environmentally. The increase in distance between facilities, while unproblematic to those with access to fast automobiles or resources to pay for more expensive modes of public transportation, puts the poor and those without access to automobiles—commonly women, children, and the infirm—at a particular disadvantage. This situation clearly raises ethical issues of social justice regarding the development of more widely dispersed communities and increased use of high-speed transportation.

Modern transportation systems clearly also raise major questions for environmental ethics. Their environmental effects cause substantial harms to individual animals and other living beings, both directly (such as by roadkill) and indirectly (by various forms of pollution), and they damage complex ecosystems and biodiversity, both directly (by infrastructural developments) and indirectly (for instance, by allowing people access to wild and remote places with increasing ease). For these ethical reasons—alongside the negative human and social effects—some writers have urged that substantial restrictions of one kind or another should be placed on access to, and growth of, modern transportation systems. (One commentator, for instance, recommends the abolition of all air transportation.) They maintain that the harms caused by such modern transportation systems outweigh their benefits. Some environmentalists also suggest that the speed and distance of modern travel means that humans become increasingly ignorant of and alienated from their natural environments, which they rush past or over in closed boxes of one kind or another. Such methods of transportation, they argue, offer no possibilities for the kinds of connections with the earth that grow out of slower, human-powered methods of travel.

Others oppose such positions, giving priority to the principle of human freedom. They maintain that the freedom to drive an automobile when and where one chooses is a personal right with which governments should not interfere. This ethical argument based on individual freedom, it is maintained, outweighs any ethical argument based on environmental harm.

This argument from human freedom has recently been attacked by Julia Meaton and David Morrice (1996). They argue that human societies do not generally allow freedom of action where substantial harms are caused to other people (for instance, we are not free to murder). Yet freedom to use automobiles *does* cause harm to humans in a variety of ways, including directly killing them in accidents. Restrictions on the use of private automobiles, therefore, would not be an unusual control on human freedom and could prevent substantial harm to humans—as well as to other animals and to the environment.

Because of the high economic, employment, and structural dependence on transportation systems in modern societies, dealing with the social and environmental harms resulting from them produces particularly difficult ethical dilemmas. As yet little work has been published by environmental ethicists that deals with issues of transportation, and it is to be expected that this will be an area of substantial growth in the future.

References: Illich, Ivan. 1974. *Energy and Equity*. London: Marion Boyars.

Meaton, Julia, and David Morrice. 1996. "The Ethics and Politics of Private Automobile Use." *Environmental Ethics* 18(1): 39–54.

Sissine, Fred. 1996. "The Partnership for a New Generation in Vehicles." Congressional Research Service Report for Congress, February 28, 96-191SPR.

United Nations Environmental Programme. 1993. *The World Environment 1972–1992*. London: UNEP/Chapman and Hall.
United States Federal Highway Administration. 1992. *1990 Nationwide Personal Transportation Survey: Summary of Travel Trends*. Washington, DC: USFHA.
Whitelegg, John. 1993. "Time Pollution." *Ecologist* 23(4): 131–134.

Waste

In general terms, waste can be defined as a substance that has been discarded or abandoned. Most legal definitions, including those used in the European Community and the United States, add that a waste may be regarded as a substance whose useful life is over, even if it has not yet been taken from the place where it was produced.

In most legal systems, waste is subdivided into a series of categories. Most important among these is hazardous waste, which in the United States can be defined as a waste that may cause or significantly contribute to mortality or serious illness because of its concentration, quantity, or physical or chemical characteristics, or may cause a substantial hazard to health or the environment if it is improperly stored, treated, transported, disposed of, or managed (Patton-Hulce 1996).

Nearly all human activities result in the production of wastes. Some wastes are, in themselves, harmless or may even be beneficial (such as, in some cases, manure and agricultural wastes). Other wastes, such as those produced by industrial and nuclear processes, may be very hazardous, causing cancers and other illnesses to human beings and animals, damaging crops and wild plants, and in the case of some toxic chemicals and radioactive wastes, causing congenital abnormalities and genetic mutations. Even relatively harmless waste, such as most domestic waste, can be problematic if it is produced in large enough amounts. It has been estimated that in the United States nearly 1,400 pounds of municipal waste is generated per person per year, of which 67 percent is dumped in environmentally destructive landfill sites (OECD 1992). Hazardous wastes create even more of a problem. The OECD estimated in 1991 that globally about 338 million tons of hazardous waste are produced per year, and that this figure is increasing. According to the OECD, a staggering 80 percent of this hazardous waste (usually composed of toxic chemicals, microorganisms, or radioactive substances) is produced in the United States. Nuclear waste is a particularly problematic element of this hazardous waste for several reasons: it is highly dangerous and can remain so for many thousands of years, there are no methods for treating it (nuclear waste cannot be treated, for instance, by incineration or the use of microorganisms), and there are no safe ways of disposing of it.

For various reasons, then, waste management is currently a growing area of concern for both the private and the public sectors. This concern has led to the promotion of a widely accepted waste management hierarchy: first, reducing the amount of waste produced; second, reusing waste substances;

third, recycling such substances; fourth, treating them (by, for instance, chemical or biological detoxification) before accepting the necessity of disposal. It should be noted, however, that waste reduction (or minimization) is usually taken to mean making production more efficient. It does not generally encompass demand management (reducing overall levels of production, thereby reducing waste generation). Given current trends in consumption, particularly in industrialized societies, overall levels of waste production are not likely to decline.

National and International Law and Policy   In the United States, the production, treatment, and disposal of waste has been a major environmental concern—in particular the disposal of hazardous (including radioactive) waste. A wide range of statutes and regulations deals with the management of waste and the cleanup of hazardous waste sites. The most important is the Comprehensive Environmental Response, Compensation and Liability Act (1980), which established the Superfund to clean up areas contaminated by hazardous waste where potentially responsible parties cannot be identified. Other important legislation includes the Resource Conservation and Recovery Act (1976, significantly amended 1984), which regulates solid and hazardous waste, controlling treatment, storage, and disposal facilities and regulating all those who generate, transportation, treat, or dispose of solid and hazardous waste.

Internationally, the disposal of hazardous wastes has also been a central environmental concern. Two agreements in particular are important: the 1972 London Dumping Convention and the 1989 Basel Convention. The first regulates the dumping of waste at sea. Annex I to the convention lists substances that may not be dumped at sea, and Annex 2 lists substances that require special care when being dumped and hence special authorization (UNEP 1993). The Basel Convention regulates the transboundary movements of hazardous wastes and their disposal. It identifies 45 categories of hazardous nonnuclear waste (nuclear waste is regulated by the International Atomic Energy Authority). Signatories to the convention affirm the right of all countries to refuse shipments of hazardous wastes from other countries; agree that all hazardous waste must be clearly labeled and handled according to legal requirements; agree not to send hazardous wastes to countries that have banned such waste or are not signatories to the convention; and recognize that if the importing country cannot dispose of the waste, the exporting country must take it back or dispose of it itself.

Relevant Ethical Issues   The production, treatment, and disposal of waste—especially hazardous waste—raises ethical questions for current and future generations of humans and for the environment, both now and in the future. At the human level, many of these ethical questions focus on equity. Does locating a potentially dangerous waste disposal site in a particular place mean that one group of people is inequitably exposed to risks while other people benefit? (Kaspersen 1983). Are local people compensated by increased employment? Are waste disposal facilities located in poorer communities or countries, where organization of protest may be weaker and the need for employment or foreign currency—regardless of the risk—stronger?

Similar questions about equity arise when the relationship between present and future generations of humans is explored. Waste is what is discarded by people who are currently alive, and some of what is discarded could be reused or recycled. Instead, it is taken out of any loop of use and reuse and discarded (incinerated or put in a landfill). This action, it might be argued, is unjust to future generations in two ways. First, present people are depleting resources (*see* Natural Resource Depletion) and, by disposing of waste that could be recycled or reused, depriving future generations of access to them. Second, the disposal of some of these wastes (for instance, in landfill sites) can damage the environment and thus affect future generations.

Where waste is hazardous or radioactive, such questions about equity are even more acute. Radioactive waste poses serious threats to human health over thousands of years. It is highly toxic in small amounts, indestructible, and its harmful effects are cumulative. No safe method exists for storing radioactive waste in the present, let alone for the long-term future. Thus, safe storage of radioactive waste is a permanent problem inflicted by the present generation on all future generations—people who may not even use nuclear power themselves and are not able to give their consent (or otherwise) to its production (Blowers, Lowry, and Solomon 1991). It produces both an economic burden, in terms of continuous monitoring and surveillance, and a health burden for the future, as containers designed to last only several hundreds of years begin to corrode. Many ethicists—from a variety of different ethical approaches—consider the production of radioactive wastes without the means for safe storage unjust and inequitable, an unacceptable way to treat future generations of human beings and reason enough to end the use of nuclear power (Schrader-Frechette 1991). Similar arguments may be made against the production of a range of other long-lasting, hazardous wastes, although few raise as many issues as radioactive wastes.

Some environmental ethicists and animal ethicists make further ethical arguments about justice here. Although most waste created by human beings (though not quite all) is generated from practices that are beneficial only to human beings, the treatment and disposal of these wastes can cause harm to animals and can damage wild ecosystems. Thus, harm and damage is caused to those who gain no benefits from the activities. It might be argued, for instance, that the location of hazardous waste disposal areas and nuclear waste stores in remote, wild areas is a way of shifting the risk from human beings, who enjoy the benefits of the technology that produces the wastes, to wild animals and ecosystems that do not enjoy the benefits, thus reflecting an implicit value hierarchy where human lives are valued more than those of other species. For some environmental philosophers, such an approach would be ethically unjust.

Of course, such ethicists would not necessarily argue that hazardous and nuclear waste sites—or even ordinary landfill sites—should be located in densely populated human areas. Instead, they would argue that *no* living beings of whatever species should be exposed to the risks generated by hazardous and radioactive waste, and that consumption patterns should change—and in some societies consumption should be reduced—so that new risks from waste are no longer being created. This argument for lifestyle changes that reduce consumption and alter consumption patterns rather than merely manage waste better is frequently advanced by environmental ethicists, in particular those adopting deep ecological positions. When effects on some present communities, future people, individuals of other species, and the environment are taken into account, it is certainly difficult to justify in ethical terms—from a range of perspectives—the volume and nature of waste production that presently characterizes industrialized societies.

References: Blowers, Andrew, David Lowry, and Barry D. Solomon. 1991. *The International Politics of Nuclear Waste*. London: Macmillan.

Kaspersen, R. 1983. *Equity Issues in Radioactive Waste Management*. Cambridge, MA: Oegelschager, Gunn and Hain.

Organization for Economic Cooperation and Development. 1992. *The State of the Environment 1991*. Paris: OECD.

Patton-Hulce, V. 1996. *Contemporary Legal Issues: Environmental Pollution*. Santa Barbara, CA: ABC-CLIO.

Schrader-Frechette, K. 1991. "Ethical Dilemmas and Radioactive Waste." *Environmental Ethics* 13(4): 327–343.

United Nations Environmental Programme. 1993. *The World Environment 1972–1992*. London: UNEP/Chapman and Hall.

Water Resources

Seventy percent of the earth's surface is water, and it is estimated that there are about 1.4 billion kilometers cubed of water on earth (Fernie and Pitkethly 1985). About 97 percent of this is saltwater in the oceans. Most of the fresh water on earth is found in the form of polar ice and snow or lies underground (UNEP 1993); only a small proportion makes up freshwater lakes and rivers. Both fresh water and ocean water are essential for the continuance of life on earth. In addition, water has a number of central uses for human beings. Water helps maintain favorable climatic patterns, provides a habitat for a variety of human food sources including fish, irrigates crops, provides a medium for transportation, can be used in energy production—both in generating hydroelectricity and in cooling of nuclear reactors—is a sink for human waste, is the site of a variety of recreational activities, provides esthetic pleasure, and in most of the above uses it can provide humans with employment.

Given the importance of water as a human resource, it is not surprising that threats to water resources are currently considered to be one of the most serious environmental problems facing human beings. The two major threats are scarcity and pollution. With an increase in human population, especially (but not only) in urban areas in developing countries, supplies of fresh water to meet drinking and sanitary needs are becoming increasingly scarce. Fresh water is equally scarce in many rural areas, where it is of central importance for irrigation. This scarcity is compounded where changing global rainfall patterns (*see* Climate Change) have meant a reduction in expected rainfall. Where fresh water is scarce and freshwater sources cross international frontiers, international tension and conflict over water extraction and use can develop. It is predicted that conflicts over water may be one of the greatest sources of international tension in coming decades.

Pollution is a major threat to all water resources, whether fresh water or marine. It may be created by deliberate discharges of waste and marine dumping, accidental leakage, or industrial and agricultural runoff. It can include, among other things, toxic or nuclear waste, oil, sewage, and agricultural fertilizers. Such water pollution can make fresh water unusable (thus exacerbating any existing scarcity) and can damage or destroy all kinds of aquatic habitats.

Water is, of course, not only a human resource. All life depends on access to it. Water scarcity and water pollution can be problematic to members of all species, as well as human beings. High levels of water extraction for human use can result in the drying out of wetlands and streams vital to the survival of other organisms, either as their drinking water or as their habitat. Pollution both of fresh water and of the sea can have similar effects (seen most dramatically after large oil spills). In addition, the construction of dams for water storage can have devastating effects on local ecosystems where dry land is flooded and also downstream, where water levels are managed and controlled.

International Law and Policy A considerable amount of international environmental law and policy exists in relation to the earth's waters. Between 1946 and 1991, 21 major international marine pollution conventions were signed. Several of these agreements are particularly significant. The International Convention for the Prevention of Pollution of the Sea by Oil (OILPOL) was first signed in 1954 and subsequently amended on a number of occasions. It was eventually superseded by the International Convention for the Prevention of Pollution from Ships (MARPOL) in 1973, which extended the area of regulation beyond the deliberate disposal of waste oil by ships at sea to include pollution from ships at sea more generally (Gorman 1993). A second agreement of key importance was the Convention on the Prevention of Marine Pollution by Dumping of Wastes and Other Matter (1972), which banned the ocean dumping of some kinds of nuclear and chemical wastes. Several other agreements concerning the protection of particular areas of ocean—including the North Sea, the Caribbean, and the Arctic—have been signed by groups of countries.

Access to and quality of supplies of fresh water have also been important issues in international policy making. In 1977, the United Nations held a conference on freshwater supplies at Mar del Plata in Argentina, which focused on achieving a "sufficient and safe supply of water for meeting global needs" (UNEP 1993). This conference led to a U.N. declaration that the 1980s should be the "international drinking water supply and sanitation decade." However, by the end of the 1980s, many of the proposed targets of this program had not been met. A further international conference on Water and the Environment held in Dublin in 1992 reviewed a number of issues concerning quality, management, and provision of water supplies and the protection of aquatic ecosystems, resulting in what was called the Dublin Statement on Water and the Environment.

Relevant Ethical Issues Water has not been widely discussed in published work in environmental ethics, perhaps because water itself is not a living organism or group of living organisms (like a species). Thus, no one is arguing that water has rights that should be respected, and water cannot suffer or be harmed.

However, the fact that water is essential to all life on earth must mean that water is of central significance to environmental ethics. The ethical issues raised concern how human use of water impacts upon other organisms, species, and living systems. For example, pollution of both fresh water and the oceans can cause pain and suffering to other organisms, especially animals and birds. It can also harm and kill aquatic plants, insects, and microorganisms; damage and destroy ecosystems; and if it is extensive enough, or the species is rare enough, it can eliminate entire species and reduce biodiversity. Likewise, excessive human water extraction can result in harm to individual organisms, species, and ecosystems.

Although there is widespread agreement about the negative effects of human water use on the nonhuman natural world, there is little agreement about how these effects should be judged ethically. Clean, fresh water is a limited resource. In many places, growing human populations are using increasing amounts of it for drinking and sanitation, agricultural and industrial purposes, and recreation. But human use threatens animals who are dependent on it for drinking, as their habitat, and to support their ecosystem. They are competitors for the same resource, and without it, they will not survive. What counts as ethical behavior in such circumstances?

Responses to this question depend on which individuals, groups, experiences, or processes are considered to be valuable. For instance, if the minimization of suffering is the basis of one's ethical position, one must weigh the human suffering caused by not increasing water extraction against the suffering caused to the other sentient beings by increasing it. It might also be helpful to ask questions about what the water is being used for—to green up a tourist golf course or to provide for fundamental human needs? Should human suffering be weighted more heavily than nonhuman suffering, in view of the greater level of psychological pain humans might endure (for instance, from being continually dirty and unable to wash)? Depending on how heavily the scales were weighted in favor of human suffering, a number of different ethical conclusions could be drawn.

But if one began from a different ethical position, arguing that the lives of all living organisms are of equal worth (a view taken, for instance, by some deep ecologists), then humans would have no ethically justifiable privileged access to water supplies. Extracting water for human use (especially for non-vital human use), where it resulted in harm or death to other organisms, would be ethically unacceptable. But if there were not enough to go around—enough for all organisms to meet their vital needs—then most of those who advocate a position of biocentric equality would accept that, as a principle one might classify as "self-defense," individual humans may ethically pursue their own survival, even at the expense of others.

However, this case raises broader political issues. The conflict here arises because of increases in human population and increasing human development (and hence, demand for water). Many deep ecologists (and some other radical environmentalists) might argue that changes in political structures and the emphasis on economic growth, reduction in population, and greater social equity would mean that such conflicts need not arise in the first place. Similar arguments might be made about marine pollution: if lifestyles less dependent on the production of toxic chemicals, on oil consumption, and on chemical fertilizers were adopted, then the conflict between harm to marine life and ecosystems and the benefits to human beings of using the sea as a dump for waste would not arise.

Clearly, access to fresh water and the effects of water pollution raise a number of difficult ethical and political questions about the distribution and use of this vital resource. As human population, urbanization, and development increases, particularly if climate change causes substantial alterations in global rainfall patterns, it seems likely that conflicts about water—among groups of humans and among humans, animals, and other elements of the natural environment—will increase. Water resources will thus inevitably become an increasingly important area of environmental work and perhaps one of the key arenas of debate in environmental ethics.

References: Fernie, John, and Alan Pitkethly. 1985. *Resources: Environment and Policy*. London: Harper and Row.

Gorman, Martha. 1993. *Environmental Hazards: Marine Pollution*. Santa Barbara, CA: ABC-CLIO.

United Nations Environmental Programme. 1993. *The World Environment 1972–1992*. London: UNEP/Chapman and Hall.

Wilderness

Defining wilderness is difficult and controversial. The term *wilderness* is commonly understood and interpreted in different ways. The most widely accepted definition is that found in Section 2c of the 1964 United States Wilderness Act, which states:

A wilderness, in contrast with those areas where man and his works dominate the landscape, is hereby recognized as an area where the earth and its community of life are untrammeled by man, where man himself is a visitor who does not remain. An area of wilderness is further defined to mean...an area of undeveloped Federal land retaining its primeval character and influence, without permanent improvements or human habitation, which is protected and managed so as to preserve its natural conditions, and which (1) generally appears to have been affected primarily by the forces of nature, with the imprint of man's work substantially unnoticeable, (2) has outstanding opportunities for solitude or a primitive and unconfined type of recreation, (3) has at least 5,000 acres, or is of sufficient size as to make practicable its preservation and use in an unimpaired condition, and (4) may also contain ecological, geological, or other features of scientific, educational or scenic value.

The first protected wilderness area in the United States was established by the U.S. Forest Service at Gila National Forest in New Mexico in 1924. The 1964 National Wilderness Act set up a National Wilderness Preservation System and gave Congress the power to designate wilderness. The act also provided for the designation of 54 wilderness areas. In 1968, the wilderness system in the United States was expanded, and by the end of 1994 there were 631 wildernesses covering 104 million acres of land in 44 states (Gorte 1994). Currently 6 percent of all land and 20 percent of all federal land is designated as, or recommended for, wilderness. The state of Alaska accounts for a significant proportion of this land: 19 percent of Alaska is wilderness, or recommended for wilderness designation, as opposed to under 4 percent of the United States as a whole (Gorte 1994).

A number of restrictions apply to the use of land designated as wilderness in the United States. It should not be used for commercial exploitation such as logging and there should be no motorized entry (except in emergencies). However, a number of exemptions do exist: some kinds of mineral prospecting are allowed, livestock grazing is permitted, and the land may be used for commercial recreation.

Relevant Ethical Issues   The idea of wilderness has, historically, been central to philosophical writing about the environment in the United States. As early as the nineteenth century, before formal designation of wilderness areas began, the writer and naturalist John Muir (1838–1914) wrote about the fundamental importance of wilderness. He argued that wilderness areas should be regarded not only as reserves of resources for human consumption but also as places for the refreshment of human spirit. Wilderness, he maintained, provides not only "fountains of lumber and irrigating rivers," but also "fountains of life" for "tired, nerve-shaken, over-civilized people" (Muir 1901). The Sierra Club, which he founded, has continued to promote this tradition. The forester Aldo Leopold (1887–1948), now regarded as a central figure in the development of environmental ethics, also promoted the protection

of wilderness as "something to be loved and cherished because it gives definition and meaning to...life" (Leopold 1949). He encouraged the Forest Service to establish the first wilderness area in 1924 and founded the Wilderness
Society in 1935. Muir and Leopold are just two figures central to the development of the idea of wilderness in the United States, a tradition that has been
explored in two important works: Roderick Nash's 1983 book, *Wilderness and
the American Mind*, and Max Oelshlaeger's 1991 book, *The Idea of Wilderness*.

The idea of wilderness has always been important in environmental ethics.
In the first volume of the journal *Environmental Ethics*, the philosopher
William Godfrey-Smith published an article in which he argued that wildernesses are valuable for human beings in a variety of ways. Like Muir, he maintained that they are places of spiritual refreshment and renewal. He also
argued that wildernesses are important scientific resources, that they protect
potentially useful biological diversity, and that they provide areas for human
recreation. Alongside these human values, Godfrey-Smith suggested that
wildernesses have nonuse or intrinsic values, and that wildernesses should be
included as part of the human "moral community."

Recently, however, wilderness has been one of the most contentious issues
in environmental ethics. Indeed, there has been considerable debate about
the very *idea* of wilderness. At the head of this debate is the philosopher
J. Baird Callicott. In 1991, Callicott published a paper in which he presents
three key arguments against the idea of wilderness. First, he argues that the
idea of wilderness in the United States is fundamentally ethnocentric,
because it ignores the fact that the land had been occupied and managed by
native Americans for thousands of years before the arrival of European set-
tlers. Second, he maintains that the idea of wilderness is a static and
unchanging idea, "pickling the land in aspic." Third, he argues that the idea
of wilderness presupposes an undesirable fundamental separation of human
beings from the land; it rests on the belief that all human alteration of pristine nature degrades it, with the practical result that wildernesses become
small temples to nature, while outside wildernesses, the destruction of the
environment continues. The idea of wilderness, he argues, "avoids facing up
to the fact that the ways and means of industrial civilization lie at the root of
the current global environmental crisis." Other criticisms of the wilderness
idea come from outside the United States. Ramachandra Guha, for example,
argues that the export of the American idea of wilderness to developing
countries in Africa and Asia has led to the physical displacement of poor
communities from their homelands and the transfer of control over the land
to rich foreign tourists (*see* Tourism).

However, these objections to the wilderness *idea*, as Callicott makes clear,
do not mean that such authors believe that wilderness areas *in practice* should
be opened to development. Particular wildernesses may in practice be important wildlife sanctuaries. Rather, both Callicott and Guha argue that humans
should focus on the idea of sustainable development where humans live in
harmony with ecosystems instead of on the idea of wilderness.

Some environmental ethicists fundamentally reject this analysis. Holmes Rolston, for instance, refutes Callicott's arguments. He maintains that many parts of the U.S. landmass currently thought of as wilderness were very little used by Native Americans, and as they are "high, cold, arid, and difficult to traverse on foot," it cannot be said that Native Americans fundamentally changed their nature (Rolston 1994). He also argues that humans are not excluded from wilderness areas; they are merely prevented from making particular kinds of uses of wilderness that would destroy the natural systems present there. Underpinning Rolston's arguments is a profound awareness of the values present in wilderness areas, including the culturally symbolic value of wilderness to Americans. But, also like Godfrey-Smith, Rolston insists that ecological systems, species, and individuals all carry intrinsic value, a value not created by human beings and not dependent on human use or appreciation (Rolston 1994). In wilderness, these values are manifested most fully; the presence of human beings in these areas would undermine the continued existence of such values. From this perspective, Rolston argues, the designation of wilderness areas is absolutely central to the protection of natural values.

These disagreements about the idea of wilderness and the relationship of human beings to wild areas have not been resolved. However, both those who argue for the protection of wildernesses as pristine areas and those who argue for sustainable development object to the ideas of groups such as the Wise Use Movement who wish to use wilderness areas for economic benefits. Thus, even though theoretical divides may persist, in practice both groups of ethicists may end up defending the same areas of land from exploitation.

References: Callicott, J. Baird. 1991. "The Wilderness Idea Revisited: The Sustainable Development Alternative." *Environmental Professional* 13: 235–247.

Godfrey-Smith, William. 1979. "The Value of Wilderness." *Environmental Ethics* 1(2): 165–171.

Gorte, Ross. 1994. "Wilderness Overview and Statistics." Congressional Research Service Report for Congress, 2 December, 94-976 ENR.

Guha, Ramachandra. 1989. "Radical American Environmentalism and Wilderness Preservation: A Third World Critique." *Environmental Ethics* 11(1): 71–83.

Leopold, Aldo. 1949. *A Sand County Almanac.* Oxford: Oxford University Press.

Muir, John. 1901. *Our National Parks.* (1992 ed.). London: Diadem Books.

Nash, Roderick. 1983. *Wilderness and the American Mind.* New Haven, CT: Yale University Press.

Oelshlaeger, Max. 1991. *The Idea of Wilderness.* New Haven, CT: Yale University Press.

Rolston, Holmes. 1994. *Conserving Natural Value.* New York: Columbia University Press.

CONTEMPORARY ETHICAL ISSUES

Chapter 5: Environmental Ethics and Environmental Law

Like English law from which it developed, there are two main sources of U.S. law: statutory law (laws enacted by a lawmaking body or legislature) and case law (laws made on reflection and interpretation of previous cases, sometimes described as law based on judicial precedent).

Sources of U.S. Environmental Law

The vast majority of U.S. statutory environmental law is made by a government administrative agency, the Environmental Protection Agency (EPA) created in 1970 by presidential executive order. Congress gave the EPA powers to create and enforce environmental regulations and to carry out inspections to ensure compliance. The EPA is not the only federal agency involved in creating environmental regulations—the Fish and Wildlife Service, for instance, designates endangered species— but it is the most important. Along with Congress and the EPA, states also pass and enforce environmental regulations, which can be tougher (but not more lenient) than federal regulations, and states may be authorized by the EPA to enforce federal regulations on their behalf within the state. If the EPA or the state decides not to pursue a violation, environmental statutes often allow citizens to pursue the matter themselves, in what are called *citizen's suits*. These have become increasingly important in the development of environmental law in the United States.

Case law has also been significant for some environmental issues in the United States. This is particularly true of the law of tort. A tort is a civil (rather than criminal) wrong where personal injury is caused by interference with an individual's protected interest. (In this sense, a company, a group of people, or some kinds of objects, such as ships, may count as legal individuals.) Torts may be deliberate or they may be caused by negligence. Torts are based on two fundamental rights: the right to prevent someone from invading one's body, and the right to prevent someone from interfering with rightfully owned property (trespassing on the person or property, for instance, or creating a nuisance toward person or property). Environmental torts can take a number of forms. If, for instance, you owned a waste incinerator that was polluting my land, I might have a case for a tort-based lawsuit for negligence. If the material that was causing the pollution was toxic, I might sue you for what is known as a *toxic tort*.

Environmental Ethics and the Law

The relationship between law and ethics in general is a complex one, and that between environmental law and environmental ethics is especially complex. One way of viewing this relationship has been suggested by law professor Christopher Stone in his book *Earth and Other Ethics* (Harper and Row, 1987). He suggests that technology provides a framework of possibility, what society is *able* to do; ethics provides a framework of morality, what societies collectively decide they *ought* to do; while law provides a tool enabling societies to implement their ethical decision making in practice.

However, this can be only the very roughest representation of the relationship between law and ethics. After all, many people hold moral principles that they would not wish to see become laws. They may oppose drinking alcohol on moral principle, for instance, but do not think Prohibition should be reinstated. Other laws rest on no direct moral principle but rather allow societies to function more smoothly (such as the law that drivers must drive on the right-hand side of the road). Furthermore, of course, questions are raised concerning the extent to which the law really does represent the collective moral choices of society. John Austin (1790–1859) argued that the law is not about morality, but is rather the command of political superiors, who can use sanctions (such as imprisonment) against those who transgress it. Marxists and anarchists have similarly argued that the law is a tool used by the more powerful in society, the ruling classes, to exercise their power over the less powerful, and that there is therefore no reason why the law should be regarded as especially ethical. However, for our purposes here, it will be assumed that, broadly speaking, the law does rest on values endorsed by the majority of those in the country it governs and provides a tool for the realization and enforcement of such values.

What does this mean for environmental ethics and law? It suggests that environmental law rests on popularly held values concerning the ways in which humans should act in their natural and living environments. But it is interesting to note that the environmental values revealed by environmental law seem to be rather different from most of those affirmed by environmental ethicists discussed in the introduction to this book. This is because nonhuman living objects (animals and plants) or natural groups or areas (ecosystems and species) do not have legal rights in the United States. In the eyes of the law, these are human property or human resources, and they are only relevant to the law where they relate to human interests as property owners or resource exploiters. Important cases related to this principle will be discussed later in this chapter, but there is a general point to be made here. The law—whether statutory or case law—is concerned not with the natural or living environment in itself but with human use of, and interests in, the environment. The law, then, reflects the anthropocentric approach to environmental ethics discussed in the introduction to this book; it does not reflect the views of many environmental ethicists concerned with values in the environment unrelated to human use. Historically, these views have been minority views in the United States, so it is not surprising that they are not reflected in the legal system. With the growth in such views, however, it is possible that changes may occur in legal approaches to animals and the natural and living environment more generally. Some environmental ethicists have already attempted to initiate change, but their success so far has been limited.

Environmental Ethics and Key U.S. Environmental Regulations

Although some federal environmental regulations existed as early as 1899, when the Rivers and Harbors Act made it illegal to dump solid waste into waterways without a permit, most environmental legislation in the United States dates from the late 1950s onward. Some of the key environmental statutes and the areas covered by them are listed below.

Atomic Energy Act 1954   Created the (now no longer existing) Atomic Energy Commission and provided for the regulation of civil nuclear power.

Air Pollution Control Act 1955   Prompted by the effects of photochemical smog in Los Angeles, this act required the U.S. Public Heath Service to research air pollution and to assist states and communities in controlling it. The Air Pollution Control Act was followed by a series of Clean Air acts, in 1963, 1967, and 1970, that provided for increased research into air quality and led to the establishment of air quality criteria.

Wilderness Act 1964 Aimed at establishing a system of wilderness preservation in the United States. It was followed by a number of acts during the 1970s and 1980s designating wilderness areas throughout the nation.

National Environmental Policy Act (NEPA) 1970 Signed by President Richard Nixon, this act was the first in a decade of increased environmental regulation.

Federal Insecticide, Fungicide and Rodenticide Act 1972 Although this act existed in an earlier form from 1949, the 1972 amendments made it into a significant piece of environmental policy. Its aim was to provide greater protection for human and environmental health while allowing continued use of insecticides, fungicides, and rodenticides.

Endangered Species Act 1973, 1982 (Amended 1976, 1977, 1978, 1979, 1980, 1988). Designed to protect rare or endangered species of plants and animals (but not bacteria, viruses, and insect pests).

Toxic Substances Control Act 1976 Controlled the introduction and use of new chemicals in the United States.

Resource Conservation and Recovery Act 1976 (Amended 1980, 1984). Essentially a revision of earlier legislation, this act primarily concerned the disposal of solid wastes.

Comprehensive Environmental Response, Compensation and Liabilities Act (CERCLA) 1980 Sometimes known as Superfund; substantially amended 1986. Established mechanisms for cleaning up heavily polluted areas. It also established the Superfund, a federal fund to assist in the costs of cleaning up pollution.

Pollution Prevention Act 1990 Intended to reduce the amount of hazardous waste produced by manufacturing industries.

All of these pieces of legislation rest on particular understandings of what is valuable and presuppose ethical approaches to environmental questions, although these presuppositions may not be obvious on first reading the legislation. To illustrate how environmental laws such as these relate to environmental ethics, we can consider the ethical basis of three of these pieces of legislation more closely: the Wilderness Act of 1964, the National Environmental Policy Act of 1970, and the Endangered Species Act of 1973 (and subsequent amendments).

Wilderness Act of 1964

Although National Parks had existed in the United States since Yellowstone National Park was established in 1872, national policy on wilderness was not crystallized until the Wilderness Act of 1964. The purpose of this act was "to establish a National Wilderness Preservation System for the permanent good of the whole people." It defines wilderness as "an area where the earth and its community of life are untrammeled by man, where man himself is a visitor who does not remain."

On the surface, there appears to be a conflict here, between wilderness as a place untrammeled by humans, who do not remain there, and wilderness as a place set aside for the permanent good of the whole people. But the underlying ethical reason for preserving wilderness here is clearly the good of human beings (specifically Americans), a good achieved by keeping the wilderness in untouched condition. This concept taps into a long-standing tradition going back to Muir and Thoreau: the belief that wild areas are places for human recreation and spiritual renewal. The act does not say that wildernesses should be protected for the sake of the animals, species, or ecosystems there, but rather for human beings.

The underlying human-centered focus of the Wilderness Act was reinforced by the fact that until 1984, the act included concessions allowing for prospecting and mining within wilderness. This human-centered understanding of environmental policy underpins all major pieces of U.S. environmental legislation, as the following two examples also indicate. The idea of wilderness as land untrammeled by human beings and deserving of protection has also been attacked recently by some environmental ethicists. (See the Wilderness entry in Chapter 4.)

National Environmental Policy Act (NEPA) 1970

This piece of environmental regulation is the cornerstone of U.S. environmental policy. It came about in response to the rise in environmental concern during the 1960s, following the publication of Rachel Carson's *Silent Spring*. Primarily, the purpose of the act was to ensure that all federal agencies took environmental protection into account in any action that might have an environmental impact, and that in such circumstances the agency should produce an Environmental Impact Statement.

Importantly for this study, the act offered a reason for the inclusion of such environmental concerns: "The Federal Government shall use all means and measures to create and maintain conditions under which man and nature can exist in productive harmony, and fulfill the social, economic and other requirements of present and future generations of Americans." Furthermore, the act included the objective of ensuring "safe, healthful, productive and esthetically and culturally pleasing surroundings for all Americans while attaining the wide range of beneficial uses of the environment without degradation, risk to health or safety, or other undesirable and unintended consequences."

Such statements clearly indicate an ethical position and reveal the values that underlie this piece of legislation. It refers to "man and nature existing in productive harmony," which suggests a philosophical position where humanity and nature are separate from one another. However, it envisions harmony between the two rather than enmity. But perhaps the key word is *productive*. Both these passages emphasize the productivity of the environment. Who is this productivity for? It is for present and future generations of Americans.

The inclusion of future generations here is important, suggesting a long-term approach to environmental questions and acknowledging ethical responsibility to people who are not yet in existence (a controversial question in ethics—*see* Sustainable Development and Future Generations in Chapter 4). However, it seems clear that environmental regulations have been introduced for the sake of Americans (by which we can safely assume the act means human Americans) rather than for the sake of animals, plants, species, or the natural and living environment more generally. As with the Wilderness Act of 1964 (see above), the environment is here viewed as a resource to be used for human benefit. So the ethical foundations of this environmental law are a range of human values: the preservation of human health and safety, human social and economic requirements, human esthetic and cultural pleasures. The environment is regarded as an instrument to achieve these. The NEPA, therefore, is based on a human-centered or *anthropocentric* approach to environmental ethics.

Endangered Species Act (ESA) 1973 (and amendments)

Although an Endangered Species Act had existed in the United States from 1966, the Endangered Species Act of 1973 is a far stronger piece of legislation than its predecessor. The 1973 act has as its stated aim the preservation of species of fish, wildlife, and plants that "are of aesthetic, ecological, educational, historical, recreational and scientific value to the Nation and its people."

It is a strongly worded act, protecting all rare or endangered animals, insects, and plants, except bacteria, viruses, and insect pests that are thought to present "an overwhelming and overriding risk to man." Endangered plant species are accorded a lesser degree of protection, although that, too, was tightened up in 1982 and 1988 amendments. The protection offered to endangered animal species includes the prevention of taking of individual members of endangered species (taking includes hunting, collecting, trapping, and killing) and also prohibits the taking of habitats if by doing so the species is threatened. Stiff penalties, including fines and imprisonment, are proposed for anyone breaking the law.

The ESA is particularly interesting from the perspective of environmental ethics. Clearly, the language used to justify species protection in the ESA is, as in NEPA, anthropocentric: the protection it provides is not for the benefit of the species, or individual members of it, but for the "Nation and its people." However, economic benefits are not listed here, as they are in NEPA. It was always recognized that the ESA might create economic costs rather than economic benefits, but human esthetic, educational, historical, ecological, and recreational values were thought to outweigh the economic costs. The recognition given to these values is, in itself, a significant ethical commitment. But the ESA also has other interesting implications. As the environmental ethicist Holmes Rolston points out in his article "Property

Rights and Endangered Species" (*University of Colorado Law Review*, vol. 61/2, 1990, pp. 283–306), the ESA fundamentally affects property rights, since it restricts what landowners can do on their land if endangered species are present. This restriction suggests that the values gained by preserving species are thought to outweigh, in part at least, the important status of property rights in the United States.

The ESA, and amendments to it, is still a controversial piece of regulation. Although the values it endorses are human-focused, they are not exclusively economic. For this reason it is a key piece of legislation for environmental ethics.

Environmental Ethics and Key U.S. Court Cases

A huge number of court cases have been underpinned by particular understandings of environmental values. They cover a wide range of issues, such as animal welfare, the development of wild areas, and water and land pollution in wild areas, especially where it damages human health and well-being and threatens endangered species. In these cases, the question arises of the legal standing of nonhuman living organisms (especially animals) and of the natural and living environment more generally. All of these are technically regarded as property or resources in the U.S. legal system. Several key court cases have clarified or developed this legal understanding of the natural world. Three very different cases will be examined here: *Sierra Club* v. *Morton* (1972), *Peck* v. *Dunn* (1978), and *Tennessee Valley Authority* v. *Hill et al.* (1978).

Sierra Club v. *Morton*, Supreme Court of the United States, no. 70-34 1972

This case concerned a proposal by Walt Disney to build a large downhill ski resort at Mineral King, a remote area of the southern Sierra Nevada owned by the U.S. Forest Service. The development, a $35 million complex covering 80 acres of the valley floor and including ski facilities on the valley sides, involved the construction of a highway through Sequoia National Park. The project was opposed by the Sierra Club.

Before 1966, organizations such as the Sierra Club had not been able to make legal challenges over developments of this kind because they were deemed to have no *economic* interest in the outcome. However, in 1966, precedent for legal challenge was set when the Second Circuit Court of Appeals upheld the legal standing of another environmental group protesting against the development of a hydroelectric project. The court ruled that the environmental group had "aesthetic, conservational, and recreational interests" in opposing the development. Following this example, in 1969, the Sierra Club brought suit in the federal district court to prevent the Forest

Service from selling the land for the Mineral King development to Walt Disney. The district court accepted the injunction, but the Forest Service appealed on the grounds that the Sierra Club did not have legal standing. The appeal was upheld by the Ninth Circuit Court of Appeals on the grounds that "the right to sue does not inure to one who does not possess it, simply because there is no one else ready and willing to assert it."

The Sierra Club then asked the Supreme Court to review the decision concerning its legal standing in the case. At this point, a key development took place, of great interest to environmental ethicists. A lawyer, Christopher Stone, entered the fray by writing an article for the *Southern California Law Review* 450 (1972) in which he argued that, rather than giving legal standing to the Sierra Club in such cases, legal standing should be conferred *upon the natural object or area itself.* The natural object should become a legal "person," possessing legal rights and able to sue for wrongs concerning it. Lawyers could act as its representative in court. Conferral of legal standing in such a way, Stone argued, was not without precedent; corporations and ships, for instance, could already be represented as "legal persons" in court. This case, thus, would be better called *Mineral King* v. *Morton*, rather than *Sierra Club* v. *Morton*, since it was Mineral King rather than the Sierra Club that would be directly harmed.

The U.S. Supreme Court delivered its decision in the case on 19 April 1972. Justice Stewart delivered the majority opinion. He maintained that the Sierra Club had not demonstrated that, as an organization, it had sufficient interest in the area to maintain injury. Though he agreed that the development might represent an "injury in fact," he maintained that the "injury in fact" test requires more than an injury to a cognizable interest. It requires that the party seeking review be himself among the injured. The case brought by the Sierra Club, he argued, did not maintain that individual members of the organization would be injured by the development, and that therefore the Sierra Club did, in this case, lack standing. The decision did suggest, however, that if the Sierra Club brought its action in different terms (that is, as a representative of a number of injured members), it might be granted legal standing.

Several judges dissented from the majority decision. One of them, Justice William O. Douglas, issued a minority opinion in which he made direct reference to Christopher Stone's article, "Should Trees Have Standing?" Accepting Stone's reasoning, he argued:

> The critical question of "standing" would be simplified and also put neatly into focus if we fashioned a federal rule which allowed environmental issues to be litigated before federal agencies or federal courts in the name of the inanimate object about to be despoiled, defaced, or invaded by roads and bulldozers and where injury is the subject of public outrage. Contemporary public con-

cern for protecting nature's ecological equilibrium should lead to the conferral of standing upon environmental objects to sue for their own preservation.

Douglas went on to argue that if the inanimate object itself has legal standing,

> there will be assurances that all the forms of life which it represents will stand before the court—the pileated woodpecker as well as the coyote and bear, the lemmings as well as the trout in the streams. These inarticulate members of the ecological group cannot speak. But those people who have so frequented the place as to know its values and wonders will be able to speak for the entire ecological community. Ecology reflects the land ethic; and as Aldo Leopold wrote in *A Sand County Almanac* (1949), "The land ethic simply enlarges the boundaries of the community to include soils, waters, plants and animals, or collectively the land." This, as I see it, is the issue of standing in the present case and controversy.

Douglas's opinion was a minority one, but like Stone's case for legal standing, it is clearly of interest for environmental ethics. The 1966 case had shown that legal interests could be established in an environmental case where human recreation or esthetic pleasure is threatened, as well as where economic interest is threatened. This decision extended the kind of environmental value that is legally recognized beyond the purely economic. However, it still depended on there being individual humans, or organizations representing individual humans, who could claim that they would be injured by the action. Stone and Douglas's proposal goes a step further, suggesting that *natural objects themselves* should be given legal standing. In this way, there would be no need for an individual human to claim injury; the object itself could be represented as an injured party.

This concept does not *necessarily* imply an ethical approach where the natural object and the living creatures associated with it are considered to be valuable independently of their usefulness to humans. However, it certainly makes such an interpretation *possible*—while the system of requiring evidence of human injury does not. Judge Douglas's comment that by conferring legal rights on natural objects we give a voice to inarticulate creatures is particularly important here. It suggests that he believes nonhuman interests should be taken into account when decisions about environmental development are made. This idea contrasts with the prevailing property-focused understanding of animals and the environment in U.S. law. For this reason—even though these concepts have not been incorporated into the law—this case was important in the development of environmental ethics and environmental law.

Peck v. *Dunn*, Supreme Court of Utah 574 P.2d 367 1978

This case, which concerned the interpretation and use of anticruelty statutes in U.S. state law, is interesting for environmental ethics primarily because of what it reveals about the ethics underpinning laws on cruelty. Salt Lake County, where the case was located, had an ordinance preventing cruelty to animals. The plaintiff, Pamela Peck, had been charged with violating this ordinance either by keeping cocks for fighting or by being a spectator at or a party to cockfighting. She brought a defensive action against the county commission, arguing that the ordinance was unconstitutional. When she lost her case, she appealed to the Supreme Court of Utah.

The decision in the case opened with some general comments on morality and on cockfighting, and it is these comments that are of particular interest here:

> It is elementary that the governing authority in the exercise of its police power has both the prerogative and the responsibility of enacting laws which will promote and conserve the good order, safety, health, morals and general welfare of society. The question is whether this ordinance is properly regarded as regulatory of morals. What constitutes morals is whatever conduct, customs and attitudes are generally accepted and approved at the time in the particular culture. It is therefore essential to consider whether cock-fighting can be regarded as merely an innocent diversion, as the plaintiff's argument seems to suggest, or is itself an evil which may be condemned by law.

> In former times the provoking and baiting of animals to fight each other for the amusement of spectators was a diversion to be accepted, or at least tolerated. Indeed, in yet earlier times, the pitting of human beings against each other as gladiators, or even against animals, was similarly viewed; and even today the fighting of animals is so accepted in some parts of the world. Over the centuries, the disposition to look on such brutalities with favor or approval has gradually lessened, and compassion and concern for man's fellow creatures of the earth has increased to the extent that it is now quite generally thought that the witnessing of animals fighting, injuring and perhaps killing one another is a cruel and barbarous practice, discordant to man's finer instincts and so offensive to his sensibilities that it is demeaning to morals. Whatever one's personal views may be of such matters, the legislative authority of our state has determined as a matter of public policy that such conduct is so involved in public morals and welfare that it has made cruelty to animals a crime and included therein the causing of one animal to fight with another.... In consequence of what we have said above, we are in agreement with the expression of learned authorities that legislation against such practices as the

fighting of animals is justified for the purpose of regulating morals and promoting the good order and general welfare of society.

First, it is notable that in this judgment the aim of the law is to safeguard "good order, safety, health, morals and general welfare of society." The society in question is clearly human society (there is surely no intention to include animals or the environment here). Thus, the law, even when it concerns cruelty to animals, does not *directly* aim to promote and conserve the safety, health, and welfare of animals. Rather, cruelty to animals is perceived as being damaging to *human* society, and for this reason particular forms of cruelty to animals, such as cockfighting, were made illegal. Gary Francione, a lawyer, discusses this point in more detail in his article "Animals, Property and Legal Welfarism: 'Unnecessary' Suffering and the 'Humane' Treatment of Animals" (46 Rutgers L. Rev. 721). He points out that the purpose of such anticruelty statutes is to improve human character rather than to protect animals. The duty not to be cruel is one owed to the whole of human society rather than to the individual animal itself.

Why is this duty regarded as a duty to human society? Some court cases, returning similar verdicts, have suggested that those who are cruel to animals may become hardened to cruelty in a way that ultimately might lead them to be cruel to humans (an argument used in the nineteenth century by the philosopher Emmanuel Kant to explain why cruelty to animals was wrong). This may have been the view of the judges in this case, or the judges may have thought that cruelty to animals was unacceptable because it distressed so many people. But the direct harm to the animals concerned is not considered to be legally relevant.

This case, then, has some relationship to the Mineral King case. Like the Mineral King natural area, animals have no direct legal standing or legal rights in U.S. law. Only human interests with respect to animals and the environment can be represented in law. Thus, legal approaches to the natural world generally differ greatly from those adopted by philosophers working in environmental and animal ethics.

Tennessee Valley Authority v. Hill et al., Supreme Court of the United States, no. 76-1701 1978

This was a key test case for the 1973 Endangered Species Act. It is of particular significance because the ESA (as passed in 1973) was regarded as such a strong piece of environmental legislation. One requirement of the act was that federal departments should "take such action necessary to ensure that actions authorized, funded, or carried out by them do not jeopardize the continued existence of such endangered species and threatened species or result in the destruction or modification of habitat of such species."

In 1975, a species of fish known as the snail darter was listed as endangered under the act. It was found only in a small area of the Little Tennessee

River—an area due to be flooded when the nearly complete Tellico Dam, then being built by the Tennessee Valley Authority, was opened. A case was then brought to prevent the completion of the dam, since this would render the species extinct. This case was turned down by the district court (on the grounds that the dam was so near completion), then reversed by the court of appeals, which argued that the act allowed for no such exceptions and suggested, "It is conceivable that the welfare of an endangered species may weigh more heavily upon the public conscience, as expressed by the final will of Congress, than the write-off of those millions of dollars already expended for Tellico in excess of its present salvageable value."

This is an interesting statement from the perspective of environmental ethicists. Although much remains unexplained (for instance, what it is about possible extinction that so troubles the public conscience), this is clearly an occasion where ethical concern about the environment might outweigh a substantial sum of public money. Furthermore, it appears to be concern about a species that is of no immediate benefit to human beings. For this reason, the statement is in itself an important one.

However, it was by no means the end of the case, which eventually passed to the Supreme Court. Chief Justice Warren Burger delivered the majority opinion in 1978, where he wrote:

> The plain intent of Congress in enacting this statute [the Endangered Species Act of 1973] was to halt and reverse the trend towards extinction, whatever the cost the legislative history reveals an explicit congressional decision to require agencies to afford first priority to the declared national policy of saving endangered species.

He continued later in his statement:

> One might [argue that] in this case the burden on the public through the loss of millions of unrecoverable dollars would greatly outweigh the loss of the snail darter. But neither the Endangered Species Act, nor article 3 of the Constitution provides federal courts with the authority to make such fine utilitarian calculations. On the contrary, the plain language of the Act, buttressed by its legislative history, shows clearly that Congress viewed the value of endangered species as "incalculable." Quite obviously it would be difficult for a court to balance the loss of a sum certain—even $100 million—against congressionally declared "incalculable" value, even assuming we had the power to engage in such a weighing process, which we emphatically do not.

This statement, important for environmental ethics, affirms that the protection of endangered species cannot be subject to a cost-benefit trade-off.

The judge makes clear that the Endangered Species Act offers absolute protection to endangered species: if their value is *incalculable*, then clearly no sum of money, however large, can outweigh their worth.

This case is particularly significant in illustrating the possible financial costs entailed by enacting and enforcing environmental legislation based on principles in environmental ethics (such as the principle of protecting endangered species). In fact, following this case, the Endangered Species Act was modified, and eventually the Tellico Dam was completed.

Conclusions

As both the environmental legislation and the court cases considered in this chapter indicate, many concerns that are central to environmental ethics are not reflected in the values endorsed by the U.S. legal system. This will remain the case as long as animals in particular and the environment more generally are classified as property.

A legal debate about a similar issue is currently taking place within the European Community, where the Treaty of Rome, the founding legal treaty of the EC, defines farm animals as "agricultural products." Environmental and animal campaigners are arguing for a change in this definition. Then these animals would be defined as sentient beings whose interests can be taken into account when European legislation is being drafted. Extensive changes of this sort (which are not likely to happen in the near future) would be required in the United States for the concerns of many environmental ethicists to be taken account of in U.S. law.

CONTEMPORARY ETHICAL ISSUES

Chapter 6:
Codes of Practice in Environmental Ethics

In recent years, codes of practice, professional codes of conduct, and ethical codes have been adopted by a growing number of organizations, industries, corporations, professions, and governments. Such codes can have a variety of functions, but they are all public commitments to particular standards of individual and/or organizational behavior. They act as an internal benchmark against which individuals or organizations may be measured to see whether their performance reaches the standards they have set for themselves.

There are very few examples of codes of practice that are explicitly called environmental ethics codes. However, several different kinds of codes of practice are relevant for environmental ethics:

1. Codes or charters of environmental responsibility created by umbrella or sector organizations that individual organizations and corporations may sign. They provide an external standard by which an organization's environmental performance can be measured. The two most important international

examples of such codes are included here: the International Chamber of Commerce's (ICC's) Business Charter for Sustainable Development, and the CERES principles composed by the Coalition for Environmentally Responsible Economies.

2. Codes of environmental good practice or environmental policies containing ethical principles adopted within organizations, both public sector organizations like universities and schools, and private sector corporations. These may take a variety of forms and appear in a variety of places: companies, for instance, may list environmental principles at the beginning of a much larger corporate environmental report.

3. Codes of professional responsibility that include short sections on professional responsibilities with respect to the environment. Two examples of such extracts are included here, from the American Institution of Architects and the Institution of Engineers, Australia.

Principles from Umbrella/Sector Organizations

The CERES principles, formerly known as the Valdez principles, were developed after the tanker *Exxon Valdez* spilled its cargo of oil in Prince William Sound, Alaska, in 1989. Since then, the principles have been revised; those reproduced here are the 1996 version. Corporations are encouraged to sign the principles as an explicit public commitment to sound environmental practices. However, few corporations have actually done so, probably because the CERES environmental charter is the most demanding of such charters in terms of the environmental standards it sets.

The ICC Business Charter for Sustainable Development, launched in 1991, grew out of existing environmental guidelines for world business. This charter has been more widely accepted by businesses and industries internationally than the CERES principles, probably because of its less stringent nature.

CERES Principles

Introduction

By adopting these Principles, we publicly affirm our belief that corporations have a responsibility for the environment, and must conduct all aspects of their business as responsible stewards of the environment by operating in a manner which protects the Earth. We believe that corporations must not compromise the ability of future generations to sustain themselves.

We will update our practices constantly in the light of advances in technology and new understandings in health and environmental science. In collaboration with CERES, we will promote a dynamic process to ensure that the Principles are interpreted in a way which accommodates changing technologies and environmental realities. We intend to make consistent, measurable progress in implementing these Principles and to apply them to all aspects of our operations throughout the world.

1. *Protection of the Biosphere:* We will reduce and make continual progress toward eliminating the release of any substance that may cause environmental damage to the air, water, or the earth and its inhabitants. We will safeguard all habitats affected by our operations and will protect open spaces and wilderness while preserving biodiversity.
2. *Sustainable Use of Natural Resources:* We will make sustainable use of renewable natural resources, such as wastes, soils and forests. We will conserve nonrenewable natural resources through efficient use and careful planning.
3. *Reduction and Disposal of Wastes:* We will reduce, and where possible eliminate possible waste through source reduction and recycling. All waste will be handled and disposed of through safe and responsible methods.
4. *Energy Conservation:* We will conserve energy and improve the energy efficiency of our internal operations and of the goods and services we sell. We will make every effort to use environmentally safe and sustainable energy sources.
5. *Risk Reduction:* We will strive to reduce the environmental, health and safety risks to our employees and the communities in which we operate through safe technologies, facilities and operating procedures, and by being prepared for emergencies.
6. *Safe Products and Services:* We will reduce and where possible eliminate the use, manufacture or sale of products and services that cause environmental damage or health or safety hazards. We will inform our customers of the environmental impacts of our products or services, and try to correct unsafe use.
7. *Environmental Restoration:* We will promptly and responsibly correct conditions we have caused that endanger health, safety or the environment. To the extent feasible, we will redress injuries we have caused to persons or damage we have caused to the environment, and will restore the environment.
8. *Informing the Public:* We will inform in a timely manner everyone who may be affected by conditions caused by our company that might endanger health, safety or the environment. We will regularly seek advice and counsel through dialogue with persons in communities near our facilities. We will not take any action against employees for reporting dangerous incidents or conditions to management or to appropriate authorities.

9. *Management Commitment:* We will implement these Principles and sustain a process that ensures that the Board of Directors and Chief Executive Officer are fully informed about pertinent environmental issues and are fully responsible for environmental policy. In selecting our Board of Directors, we will consider demonstrated environmental commitment as a factor.

10. *Audits and Reports:* We will conduct an annual self-evaluation of our progress in implementing these Principles. We will support the timely creation of generally accepted environmental audit procedures. We will annually complete the CERES Report, which will be made available to the public.

11. *Disclaimer:* These Principles establish an environmental ethic with criteria by which investors and others can assess the environmental performance of companies. Companies that endorse these principles pledge to go voluntarily beyond the requirements of the law. The terms may and might in Principles one and eight are not meant to cover every imaginable consequence, no matter how remote. Rather, these Principles obligate endorsers to behave as prudent persons who are not governed by conflicting interests and who possess a strong commitment to environmental excellence and to human health and safety. These Principles are not intended to create new legal liabilities, expand existing rights or obligations, waive legal defenses, or otherwise affect the legal position of any endorsing company, and are not intended to be used against an endorser in any legal proceeding for any purpose.

Discussion

These principles, which explicitly describe themselves as prescribing an environmental ethic, emphasize the importance of corporate responsibility for the environment. The fundamental philosophical position they reflect is environmental stewardship, which seems to be defined as looking after the environment on behalf of future generations. In fact, the introductory section to the principles suggests that the reason for environmental protection is indeed the benefit of future generations. There is no argument that the environment might have value in itself.

The CERES principles commit the signing organization to ethical behavior toward the environment in a variety of contexts; the protection of wilderness, biodiversity, human health, and scarce resources are all explicitly mentioned. In addition, the principles commit companies to opening to the public their records about their environmental affairs and any environmental hazards they may be creating. Thus, the principles link an environmental ethic with the idea of freedom of information.

The ICC Business Charter for Sustainable Development

1. *Corporate Priority:* To recognise environmental management as among the highest corporate priorities and as a key determinant to sustainable development; to establish policies, programmes and practices for conducting operations in an environmentally sound manner.
2. *Integrated Management:* To integrate these policies, programmes and practices fully into each business as an essential element of management in all its functions.
3. *Process of Improvement:* To continue to improve corporate policies, programmes and environmental performance, taking into account technical developments, scientific understanding, consumer needs and community expectations, with legal regulations as a starting point; and to apply the same environmental criteria internationally.
4. *Employee Education:* To educate, train and motivate employees to conduct their activities in an environmentally responsible manner.
5. *Prior Assessment:* To assess environmental impacts before starting a new activity or project and before decommissioning a facility or leaving a site.
6. *Products and Services:* To develop or provide products or services which have no undue environmental impact and are safe in their intended use, that are efficient in their consumption of energy and natural resources, and that can be recycled, reused and disposed of safely.
7. *Customer Advice:* To advise, and where relevant educate, customers, distributors and the public in the safe use, transportation, storage and disposal of products provided; and to apply similar consideration in the provision of services.
8. *Facilities and Operations:* To develop, design and operate facilities and conduct activities taking into consideration the efficient use of energy and materials, the sustainable use of renewable resources, the minimisation of adverse environmental impact and waste generation, and the safe disposal of residual wastes.
9. *Research:* To conduct or support research on the environmental impacts of raw materials, products, processes, emissions and wastes, associated with the enterprise and on the means of minimizing such adverse impacts.
10. *Precautionary Approach:* To modify the manufacture, marketing or use of products or services, or the conduct of activities, consistent with scientific or technical understanding, to prevent serious or irreversible environmental degradation.

129

11. *Contractors and Suppliers:* To promote the adoption of these principles by contractors acting on behalf of the enterprise, encourage, and where appropriate, requiring improvements in their practices to make them consistent with those of the enterprise; and to encourage the wider adoption of these principles by suppliers.

12. *Emergency Preparedness:* To develop or maintain, where significant hazards exist, emergency preparedness plans in conjunction with emergency services, relevant authorities and local communities, recognizing potential transboundary impacts.

13. *Transfer of Technology:* To contribute to the transfer of environmentally sound technology and management methods throughout the industrial and public sectors.

14. *Contributing to the Common Effort:* To contribute to the development of public policy and to business, governmental and intergovernmental programmes and educational initiatives that will enhance environmental awareness and protection.

15. *Openness to Concerns:* To foster openness and dialogue with employees and the public, anticipating and responding to their concerns about the potential hazards and impacts of operations, products, wastes or services, including those of transboundary or global significance.

16. *Compliance and Reporting:* To measure environmental performance; to conduct regular environmental audits and assessments of compliance with company requirements, legal requirements and these principles; and periodically, to provide appropriate information to the Board of Directors, shareholders, employees, the authorities and the public.

Discussion

The ICC Charter focuses primarily on the development of policies, management, and processes that contribute to sustainable development and to environmental soundness. Clearly, such a commitment must rest on some kind of ethical conviction, but commitment to an environmental ethic is not explicitly stated. Assuming that "sustainable development" lies at the heart of this charter, we can assume that, like the CERES principles, the well-being of future generations of human beings (rather than ethical concern for the environment independent of its use value) is the primary motivation here.

The ICC Charter does not mention specific commitments (such as to wilderness protection), which are endorsed by the CERES principles. In general, the CERES principles expect a much stronger commitment on the part of signatories. The CERES principles also talk about moving toward the *elimination* of environmentally damaging substances and wastes; the ICC

Charter confines itself to discussion of "minimization" and "safe disposal." The CERES principles also commit signatories to a much greater degree of environmental disclosure—a point that is rather vague in the ICC Charter.

Further Umbrella/Sector Codes

Although the CERES and ICC codes are the two main sets of environmental principles for businesses and industry, a variety of others exist for industry and also for public sector organizations. The chemical industry, for instance, has a set of guiding principles called the Chemical Industry Responsible Care Program, which companies in that industrial sector may sign. Religious organizations have also produced sets of environmental principles that corporations may sign. For instance, the Ecumenical Committee for Corporate Responsibility, together with the Interfaith Center on Corporate Responsibility and the Taskforce on the Churches and Corporate Responsibility, produced a set of Principles for Global Corporate Responsibility: Benchmarks for Measuring Business Performance in 1995. Examples of public sector ethical codes are the Taillores Principles and the Barnabus Principles, two key sets of principles that educational organizations may sign. In the United Kingdom, the environmental organization Friends of the Earth established an Environmental Charter for Local Government, which local authorities could sign, that included 15 explicit environmental policy commitments.

Organizational Ethical Codes

Alongside environmental charters or ethical codes that a number of organizations may sign are individual environmental codes adopted by organizations for their own internal regulation. These may take a variety of forms, the most common of which is an organizational environmental policy. It is based on an ethical premise and includes a number of commitments. In the case of businesses, these environmental policies or codes of practice may appear within an annual environmental report. For all organizations that adopt them, they should form a point of reference for all policy making in the organization and may result in the constraining of activities that the organization might otherwise have undertaken. Some corporations include acceptance of their ethical codes as part of their contract of employment. Examples of such environmental policies follow.

Volkswagen Environmental Policy—Basic Principles

1. It is the declared aim of Volkswagen in all its activities to restrict the environmental impact to a minimum and to make its own contribution to resolving environmental problems at a regional and global level.

2. It is Volkswagen's aim to offer high-quality automobiles which take equal account of the expectations of its customers with regard to environmental compatibility, economy, safety, quality and comfort.

3. In order to safeguard the long-term future of the company and enhance its competitive position Volkswagen is researching into and developing ecologically efficient products, processes and concepts for personal mobility.

4. Those responsible for environmental management at Volkswagen shall, on the basis of the company's environmental policy, ensure that in conjunction with suppliers, service providers, retailers and recycling companies, the environmental compatibility of its vehicles and production plants is subject to a process of continual improvement.

5. The Volkswagen Board of Management shall, at regular intervals, check that the company's environmental policy and objectives are being observed and that the Environmental Management System is working properly. This shall include evaluation of the recorded environmentally relevant data.

6. Providing frank and clear information and entering into dialogue with customers, dealers and the public is a matter of course for Volkswagen. Co-operation with policy-makers and the authorities is based on a fundamentally proactive approach founded on mutual trust and includes provision for emergencies at each production site.

7. In keeping with their duties, all Volkswagen employees are informed, trained and motivated in respect of environmental protection. They are under obligation to implement these principles and to comply with statutory provisions as these apply to their respective activities.

(Extract from pp. 12–13, the *Volkswagen Environmental Report 1996*)

Discussion

Volkswagen's "basic principles" are interesting because of the general lack of ethical commitment they entail. Three of the principles (3, 6, and 7) are descriptive, telling the reader what Volkswagen is currently doing rather than affirming an aim, or standard, to which the company pledges to adhere. In this respect, it is odd that they are called principles at all. In addition, the third principle makes it clear that ecologically efficient technology is being developed "[i]n order to safeguard the long-term future of the company and enhance its competitive position." This clearly reflects corporate interest rather than an ethical commitment. Principle 2 suggests only that Volkswagen considers "environmental compatibility" (an odd expression that remains unexplained) alongside other elements in automobile manufacture (including quality and comfort, which might, after all, conflict with environmental concerns). Principles 4 and 5 make some real commitments to environmental improvement, but the nature and degree of this commitment remains unspecified. Only Principle 1 here has any force, in maintaining that the company is resolved to restrict its environmental impact to a minimum and to contribute to the resolution of regional and global environmental problems. Even this commitment is rather vague, and it is not at all clear what its underlying ethical basis might be.

IBM Corporate Environmental Policy

IBM is committed to environmental affairs leadership in all of its business activities. IBM has long-standing corporate policies of providing a safe and healthful workplace and safe products (Policy Letter Number 127), protecting the environment (Number 129), and conserving energy and natural resources (Number 131) which were initiated in 1967, 1971 and 1974 respectively. These policies continue to guide our operations and they are the foundation for the following corporate policy objectives:

- Provide a safe and healthful workplace, including avoiding or correcting hazards and ensuring that personnel are properly trained and have appropriate safety and emergency equipment.

- Be an environmentally responsible neighbor in the communities where we operate and act promptly and responsibly to correct incidents or conditions that endanger health, safety, or the environment, report them to the authorities promptly, and inform everyone who may be affected by them.

- Maintain respect for natural resources by practicing conservation and striving to recycle materials, purchase recycled materials and use recyclable packaging and other materials.

- Develop, manufacture and make products that are safe in their intended use, efficient in their use of energy, protective of the environment, and that can be recycled or disposed of safely.

- Use development and manufacturing processes that do not adversely affect the environment, including developing and improving operations and technologies to minimize waste, prevent air, water and other pollution, minimize health and safety risks and dispose of waste safely and responsibly.

- Ensure the responsible use of energy throughout our business, including conserving energy, improving energy efficiency, looking for safer energy resources and giving preference to renewable over non-renewable energy sources where feasible.

- Assist in the development of technological solutions to global environment problems, share appropriate pollution prevention technology and methods, and participate in efforts to improve environmental protection and understanding throughout the industry.

- Meet or exceed all governmental requirements. Where none exist, set and adhere to stringent standards of our own, and continually improve these standards in the light of technological advances and new environmental data.

- Conduct rigorous audits and self-assessment of IBM's compliance with this policy, measure progress of IBM's environmental affairs performance, and report periodically to the Board of Directors. Every employee and every contractor on IBM premises is expected to follow the company's policies and to report any environmental, health, or safety concern to IBM Management. Managers are expected to take prompt action.

Corporate Policy no. 139 from *IBM: 20 Years of Commitment*, IBM Environmental Programmes, UK, 1995.

Discussion

IBM's corporate environmental policy is clearly more detailed than that of Volkswagen. The text suggests a number of ethical reasons why an environmental policy might be necessary: human health and safety, respect for natural resources, and environmental protection. However, it does not indicate whether "environmental protection" is considered important for its own sake or only inasmuch as the environment is a human resource (perhaps for the use of future generations). Many of the commitments are rather vague—"striving" to recycle materials, for instance, or using energy "responsibly." However, IBM's policy does cover a wide range of operational areas: energy, waste, purchasing policy, nature of final product, and effect on other industries in the sector. For these reasons, IBM's policy is regarded as one of the more rigorous existing environmental policies in industry.

University of Edinburgh
Environmental Policy Statement

The University of Edinburgh recognizes that its activities impact upon the environment at local, regional and global levels and acknowledges a responsibility for the protection of the environment and of the health of its members and the wider community.

The University is committed to:

- Promoting the protection of the environment, and minimizing the impact of all its activities upon each of the local, regional and global environments both directly and through its influence on others

- Integrating environmental management policies and practices into every level and every department of the University

- Providing safe, healthy working conditions for staff and students

- Contributing to a sustainable and healthy future by conserving natural resources, by minimizing avoidable waste and pollution and by reducing and discouraging litter, graffiti and noise pollution

- Reducing the use of fossil fuels through improvements to energy efficiency and the substitution of renewable energy sources

- Avoiding the unnecessary use of hazardous materials and processes and taking all reasonable steps to prevent damage to either public or ecological health where such materials are in essential use

- Developing effective waste management and recycling procedures and using recycled and recyclable material where possible

- Establishing a rationalized transport policy; encouraging the use of public transport and providing improved facilities for the disabled, pedestrians and cyclists

- Protecting natural habitats and local wildlife and preserving biological diversity

- Increasing awareness of environmental responsibilities among staff and students through staff development and training and through initiatives in Environmental Teaching and Research.

Environmental Policy Statement. Reproduced from pp. 1–2, *The University's Agenda for the Environment*, University of Edinburgh, UK, 1993.

Discussion

This environmental policy, formally adopted in 1993, stems from a public educational institution, the University of Edinburgh, U.K., rather than from a corporate, shareholding body. It recognizes the wide range of environmental impacts an institution as large as a university may have on local, regional, and global environments, and makes some degree of commitment to lessen these impacts. Some of the ethical reasons for making such an environmental commitment are clear: for instance, in maintaining the health and safety of employees. An ethical commitment to "the future" is also expressed; since this commitment is placed in the context of conserving natural resources and discouraging litter and graffiti, it seems likely that the focus is on the future of human beings rather than the environment, or elements of it, itself. Although the need for environmental protection and the preservation of biological diversity is stated, the policy does not make the reasons clear. Are these valuable goals in themselves, or is the environment worth preserving because it is useful to present and future human beings?

Despite its lack of ethical clarity, this institutional policy is generally more rigorous than the corporate policies included earlier. Although the language of reduction and minimization of resource use is imprecise, the policy does cover a wide range of operational areas. It includes, for example, a commitment to promoting public transport, which is absent from most corporate environmental policies (not surprising, perhaps, in the case of Volkswagen). This greater degree of policy rigor is to be expected, considering the aims, "customers," and employees of an educational institution in comparison with those of a corporation. Many educational institutions, in particular institutions of higher education, have adopted environmental policies or codes of ethics like those of the University of Edinburgh. (Details of many of these, as well as the umbrella charters educational institutions can sign, can be found at the website for the International Institute for Sustainable Development in Canada: http://iisd1.iisd.ca/.)

Codes of Professional Ethics

Codes of professional ethics are established by organizations representing groups of individuals who practice particular professions—for example, doctors, lawyers, engineers, teachers. These professionals are often expected to abide by the codes of ethics adopted by their professional body in order to retain their professional accreditation. Such codes of ethics may cover wide areas, including topics such as how professionals should deal with their clients, what kinds of professional confidentiality they should adhere to, what sorts of relationships they should have with other professionals, and what sorts of obligations they should have to the human community at large.

Although many of these professional codes of ethics are closely concerned with matters of human health and safety, surprisingly few (in the United States at least) address environmental concerns. Those that do include some mention of the environment usually pass over it in a few sentences, although the code is many times longer. But a few professional organizations, especially in areas where members might be perceived to have particular environmental impact, have drawn up more detailed environmental principles. Examples of both are reprinted below.

The American Institute of Architects

The American Institute of Architects revised its Code of Ethics and Professional Conduct in 1993. Its code has three classes: Canons, Ethical Standards, and Rules of Conduct. Canons are understood to be broad principles, Ethical Standards are specific goals, and Rules of Conduct are obligatory—violation of these leads to disciplinary action. The environmental section is listed as an Ethical Standard and is buried in the middle of a long document. It reads:

> **E.S. 2.2 Natural and Cultural Heritage:** Members should respect and help conserve their natural and cultural heritage while striving to improve the environment and the quality of life within it.

(Gorlin, Rena, ed. *Codes of Professional Responsibility*, 3d ed., Bureau of National Affairs, Inc., 1994.)

Not only is this ethical standard brief, it is unclear. Does the description of the environment as "cultural and natural heritage" suggest that the importance of the environment is understood here in terms of human inheritance? The improvement of the environment may be desirable in a rundown urban area, but what might it mean in a wilderness? Does quality of life mean quality of human life or quality of life for all that is living? This commitment is too brief to really have any ethical clout for those in the profession concerned.

Institution of Engineers, Australia

In contrast with the American Institute of Architects, the Institution of Engineers, Australia, issued a detailed set of *Environmental Principles for Engineers*, prepared by the National Committee on Environmental Engineering in 1992. This document, which is several pages long and too substantial to include in its entirety, takes specific account of what it calls a "sustainability ethic." The ethic is based on the following three principles:

1.1 Recognize that ecosystem interdependence and diversity form the basis for our continued existence.
1.2 Recognize the finite capacity of the environment to assimilate human made changes.
1.3 Recognize the rights of future generations. No generation should increase its wealth to the detriment of others.

(Taken from *Environmental Principles for Engineers*, p. 1, Institution for Engineers, Australia, 1992)

These three foundational principles are unusual in professional codes of ethics, both in their clarity and in the strength of the language used. (It is especially unusual, for instance, to see an affirmation of the rights of future generations.) However, the ethical concern here is very clearly for human beings, both present and future. The environment is understood to be a *resource* that must be shared fairly between generations. This understanding is repeated throughout the document, which prescribes various kinds of behavior that conform to the recognition of these basic principles. This behavior includes consuming the smallest possible amount of raw materials and energy; urging clients to incorporate environmental objectives into design; and considering the "consequences of all proposals and actions, direct or indirect, immediate or long term, upon cultural heritage, social stability, health of people and equity." These are, then, a wide-ranging set of professional ethics principles. Although their motivation is the protection of human health and resources in the present and future, the implementation of such principles would clearly have beneficial long-term environmental impacts as well.

Conclusions

In principle, ethics codes, principles, and charters are important and helpful; in practice, their relationship with environmental ethics, and with the environmental movement in general, is uneasy. One reason for this is that nearly all existing ethics codes have sidestepped the question (which would, for instance, be pressed by deep ecologists) of whether the natural world, or at least some elements of it, has value independent of human use. All of the codes discussed here urge environmental conservation because it is to the short- and long-term benefit of human beings, not because it is beneficial for the natural environment itself. In response, it may be argued that the idea that the natural environment is valuable apart from its resource value to humans is a controversial one that need not be dragged into such practical documents as ethics codes. This argument would be strengthened if it could be shown that the same degree of conservation could be achieved for human-centered reasons as for nature-centered reasons.

However, this is not the only objection that has been voiced to some kinds of environmental ethics codes and charters in practice. Where such codes are adopted by corporations, particularly corporations whose business appears to be by nature environmentally damaging (such as mining corporations, oil companies, and automobile manufacturers), they are sometimes regarded as "greenwash." In other words, it may be argued that the company adopts the code to create the appearance of environmental concern without any fundamental change in previous destructive policies. It can then claim to an environmentally concerned public that it has undertaken a process of organizational "greening," yet make few changes in its organizational practice. Thus, an environmental ethics code might simply be part of corporate public relations rather than a genuine attempt to introduce ethical principles into corporate environmental activities.

Clearly, it is impossible to generalize about the nature of, and motivation behind, corporate adoption of environmental policies and principles. A report by the U.K. Association of Certified and Chartered Accountants found that the majority of businesses in Europe that had adopted environmental policies said that they had done so for the good of the company rather than for ethical reasons (as Volkswagen, above). (Adams, Hill, and Roberts, *Environmental Employee and Ethical Reporting in Europe*, ACCA Research Report 41, 1994). If this is so, then it is likely that where the overall good of the company could be increased by having weak environmental policies—or ignoring policies that are in place—the company will pursue such a path. The lack of mechanisms external to the organization for policing such environmental pledges must surely add to the temptation to disregard them.

The same questions of corporate interest do not arise, however, in relation to professional codes of ethics. It is surprising that more consideration of environmental ethics has not, to date, entered codes of professional responsibility. Perhaps environmental ethics will expand in this area in the future.

CONTEMPORARY ETHICAL ISSUES

Chapter 7: Annotated Directory of Organizations

There is a wide range of organizations and associations whose work is relevant and important to environmental ethics. They fall into four main categories: academic centers, with an emphasis on research and teaching; environmental pressure groups, which campaign on issues relating to environmental ethics; federal and state agencies; and organizations, both national and international, of professionals who are concerned with environmental ethics.

Academic Centers for Environmental Ethics

The selection listed below includes the largest and best-known academic departments and centers that work in environmental ethics. The web page at the University of North Texas (http://www.cep.unt.edu) provides a detailed and up-to-date summary of academic centers in the field.

Bowling Green State University
Department of Philosophy
Bowling Green, OH 43403
(429) 372-2086
Website: http://www.bgsu.edu/departments/phil/program

Home of the Social Philosophy and Policy Center, the philosophy department specializes in applied ethics, value theory, and moral and political philosophy. Graduate students can earn a Ph.D. in applied philosophy with a concentration in environmental philosophy. Students are encouraged to take internships in environmental fields.

Publications: *Social Philosophy and Policy* (quarterly journal).

Center for Development and the Environment
University of Oslo
P.O. Box 1116, Blindern
N-0317 Oslo, Norway
47 22 85 89 00
Fax: 47 22 85 89 20
Website: http://www.sum.uio.no

Part of the University of Oslo, this center is dedicated to the generation and dissemination of knowledge within the field of environment and development. It researches areas such as social conflict, poverty, and the environmental crisis and sustainable development.

Publications: Variety of publications available free of charge to libraries.

Center for Environmental Philosophy
University of North Texas
P.O. Box 13496
Denton, TX 76203-6496
(817) 565-2727
Fax: (817) 565-4448
E-mail: ee@unt.edu
Website: http://www.cep.unt.edu
Director: Eugene Hargrove

This center was established in 1989 to facilitate teaching and research in environmental ethics. It became an affiliated organization of the University of North Texas in 1991. Its major activities include the reprinting of important books in environmental ethics, holding workshops and conferences, and graduate, postdoctoral, and professional education. Eugene Hargrove, the director of the center, is also the editor of the journal *Environmental Ethics*; also on the board of directors are J. Baird Callicott, Max Oelshlaeger,

Holmes Rolston, and Thomas Birch. The Center for Environmental Philosophy offers master's degrees in philosophy with a concentration on environmental ethics, maintains a website on environmental ethics, and sponsors a number of visiting researchers in environmental ethics.

Publications: *Environmental Ethics* (quarterly journal).

Center for the Study of Values in Public Life
Harvard University
56 Francis Avenue
Cambridge, MA 02138
(617) 496-5208
Website: http://divweb.harvard.edu/csvpl.ee

This center runs an environmental ethics and public policy research program, a seminar on environmental values, and a faculty-student discussion group.

Publications: A series of occasional papers and subject bibliographies on environmental ethics.

Centre for Philosophy and the Environment
University of Manchester
Department of Philosophy
Manchester M13 9PL
UK
44 (0)161 275 3196
Fax: 44 (0)161 275 3613
E-mail: mfestkkl@fsl.man.ac.uk
Director: Keekok Lee

This center runs an interdisciplinary master's degree program in philosophy and the environment, including core courses in environmental philosophy and options in a range of other topics including environmental economics and biology.

Colorado State University
Department of Philosophy
Fort Collins, CO 80523
(970) 491-6315
Fax: (970) 491-4900
Website: http://www.cep.unt.edu/colorado.html

Offers undergraduate and graduate classes in environmental ethics but has no Ph.D. program. Holmes Rolston, one of the first philosophers to work in environmental ethics, is a professor here.

Department of Philosophy
Furness College
University of Lancaster
Lancaster LA1 4YG
UK
44 (0)1524 592 490
Fax: 44 (0)1524 846 102
E-mail: philosophy@lancaster.ac.uk
Website: http://www.lancs.ac.uk/users/philosophy

This is the main center for work in environmental ethics in the United Kingdom. The department offers a master's program (founded in 1988) in Values and the Environment, which includes units on ethical theory and the environment, biotechnology and science, and also runs weekly open seminars on issues in environmental ethics.

Publications: *Environmental Values* (quarterly journal).

The George Perkins Marsh Institute
P.O. Box 1
Viola, ID 83872-0001
(208) 883-0626
E-mail: witt731@uidaho.edu
Website: http://www.uidaho.edu/~witt731/gpmi.html

Founded on the original Earth Day in 1970, the George Perkins Marsh Institute was formed to promote the ecological foundations of human communities. It is constituted by a loose grouping of ecologists working on independently designed and funded projects based on holistic and ethically and environmentally sensitive approaches to science.

Publications: *Pan Ecology* (irregularly produced journal).

Keele Centre for Environmental Research
University of Keele
Keele, Staffordshire ST5 5BG
UK
E-mail: ira13@cc.keele.ac.uk

This U.K. research center works in a wide range of environmental areas, in particular in environmental politics and public policy. The center runs a master's degree program in environmental politics, which includes courses relevant to environmental ethics and philosophy.

McGregor School of Antioch University
Environment and Community Program
800 Livermore Street

Yellow Springs, OH 45387
(513) 767-6321
Website: http://www.cep.unt.edu/antioch.html

This program offers an M.A. in environment and community; environmental philosophy is included among the core courses.

Oxford Centre for Environment, Ethics and Society (OCEES)
Mansfield College
Oxford OX1 3TF
UK
Tel/Fax: 44 (0)1865 270886
E-mail: ocees@mansfield.ox.ac.uk.
Website: http://users.ox.ac.uk/~ocees/

This center is a multidisciplinary research institute. Founded in 1992, it is dedicated to exploring the social and ethical aspects of environmental questions with a multidisciplinary research team.

Publications: Series of occasional papers, some of which explicitly concern issues in environmental ethics.

The Sierra Institute
Box AA
University of California Extension
740 Front Street #155
Santa Cruz, CA 95000
(408) 427-6618

This institute offers a course in Wilderness Field Studies, which includes units in environmental ethics and the philosophy of nature.

University of Georgia
Environmental Ethics Certificate Program (EECP)
Department of Crop and Soil Sciences
3111 Plant Sciences
Athens, GA 30602-7272
(706) 542-0898
Fax: (706) 542-0914
Website: http://www.phil.uga.edu/eecp/

This is an interdisciplinary certificate program providing a forum for discussion about environmental problems that involve competing values. The program offers undergraduate and graduate certificates (which include compulsory work in environmental ethics) as well as a series of evening seminars, occasional conferences, and "philosophers' walks."

Publications: *Ethics and the Environment* (quarterly journal).

University of Montana
Department of Philosophy
Missoula, MT 59812
(406) 243-4076
Fax: (406) 243-2171
Website: http://www.cep.unt.edu/other/montana.html

During 1996, the philosophy department launched a master's course in environmental philosophy. Students can specialize in a range of areas including environmental ethics, ecofeminism, wilderness, environment and agriculture, and environment and technology.

University of Stellenbosch
Department of Philosophy
7600 Stellenbosch, South Africa
27 21 808 2058
Fax: 27 21 886 4343
E-mail: jph2@maties.sun.ac.za.
Website: http://www.cep.unt.edu/stellen.html
Director: John Hattingh

Has a Unit for Environmental Ethics, which offers M.A. and Ph.D. work and other professional training. The focus is on environmental ethics in contemporary society, particularly the environmental problems facing South Africa.

Worldwatch Institute
1776 Massachusetts Avenue NW
Washington, DC 20036
(202) 452-1999
Fax: (202) 296-7365
E-mail: wwpub@igc.apc.org
Website: http://www.worldwatch.org

The Worldwatch Institute is dedicated to fostering a sustainable society. It publishes interdisciplinary research on global issues to achieve this end.

Publications: *State of the World*, an annual guide for government officials, environmentalists, and students. A wide range of other environmental publications is available.

Environmental Special Interest Groups

Nearly all environmental special interest groups are driven by ethical convictions of one kind or another. It is only possible to include a small selection of such groups here, focusing on those who are especially interested in environmental ethics issues.

Earth First!

P.O. Box 1415
Eugene, OR 97440
(503) 741-9191
Fax: (503) 741-9192
E-mail: earthfirst@igc.apc.org
Website: gopher://gopher.igc.apc.org/11/orgs/ef.journal

Earth First! is a radical environmental group with the slogan "No Compromise in Defense of Mother Earth!" It advocates direct action on environmental issues—in particular, wilderness development—including civil disobedience and sabotage ("monkeywrenching"). It does not have formal members, but relies on the support of activists.

Publications: The *Earth First!* journal, published eight times a year on pagan holidays.

Environmental Defense Fund

257 Park Avenue South
New York, NY 10010
(212) 505-2100
Website: http://www.edf.org

Founded in 1967, the Environmental Defense Fund is well known for its lawsuits on environmental issues (for instance, in banning the pesticide DDT). Employees include lawyers, scientists, and economists, and recent activities include the promotion of economic instruments in environmental policy.

Friends of the Earth

1025 Vermont Avenue NW, Suite 300
Washington, DC 20005-6303
(202) 783-7400
Fax: (202) 783-0444
Website: http://www.essential.org/FOE/FOE.html

Friends of the Earth is a nonprofit advocacy organization dedicated to protecting the planet from environmental degradation, preserving diversity (biological, cultural, and ethnic), and empowering citizens politically.

Greenpeace USA

1436 U Street NW
Washington, DC 20009
(202) 462-1777
Fax: (202) 462-4507
Website: http://www.greenpeace.org

An international campaigning organization, Greenpeace aims to protect bio-diversity, prevent pollution, end nuclear threats, and promote peace, primarily by methods of nonviolent confrontation.

National Audubon Society

700 Broadway
New York, NY 10003-9501
(212) 979-3000
Website: http://www.audubon.org/audubon/

Founded in 1905, the National Audubon Society is an important U.S. conservation organization. Its mission is to conserve and restore natural ecosystems, focusing on birds and other wildlife, for the benefit of humanity and the earth's biological diversity. It runs a number of nature reserves.

Natural Resources Defense Council (NRDC)

40 West 20th Street
New York, NY 10011
(212) 727-2700
E-mail: nrdcinfo@nrdc.org
Website: http://www.nrdc.org

In its mission statement, the NRDC emphasizes its ethical commitment to sustainability and good stewardship of the earth. It aims to "safeguard the Earth: its people, its plants and animals and the natural systems on which all life depends." It is also concerned about the environmental welfare of the disadvantaged and future generations.

People for the Ethical Treatment of Animals (PETA) US

501 Front Street
Norfolk, VA 23510
(757) 622-7381
Fax: (757) 622-1078

PETA claims to be the largest animal rights organization in the world, and its mission is to establish and protect the rights of all animals. It maintains that "animals are not ours to eat, wear, experiment on or use for entertainment." PETA concentrates its work on factory farms, laboratories, the animal entertainment industry, and the fur trade. It undertakes educational, campaigning, and investigative work.

Sierra Club

730 Polk Street
San Francisco, CA 94109
(415) 776-2211
Fax: (415) 776-0350
Website: http://www.sierraclub.org

Founded in 1892 with John Muir as its first president, the Sierra Club is a nonprofit public interest organization. It works to promote conservation of the environment by influencing public policy in a variety of environmental areas, including protection of wilderness and environmental education.

Publications: The Sierra Club publishes *Sierra Magazine* and a wide range of newsletters and books, including *The Sierra Club Green Guide: Everybody's Desk Reference to Environmental Information*.

World Wildlife Fund
(Known as the Worldwide Fund for Nature except in the
United States and Canada)
1250 Twenty-Fourth Street NW
P.O. Box 96555
Washington, DC 20037
(202) 293-4800
Website: http://www.wwf.org

Founded in 1961, the World Wildlife Fund is the world's largest independent conservation organization, with over 4.7 million supporters. It aims to preserve biodiversity at all levels (genetic, species, and ecosystem), to ensure sustainable use of resources, and to minimize pollution. In the United States, it has lobbied for the implementation of the Endangered Species Act. It carries out practical conservation work but also has an education program to develop understanding of environmental problems and issues relating to environmental ethics.

Publications: *Focus* is a bimonthly newsletter for members.

Federal Agencies

There are a number of federal agencies with responsibility for different areas of federal environmental policy. The most important are listed below.

Environmental Protection Agency (EPA)
401 M Street SW
Washington, DC 20024
(202) 260-2090
Website: http://www.epa.gov/

The Environmental Protection Agency was founded in 1970, after the passing of the Environmental Protection Act. It creates (subject to congressional approval) and enforces U.S. environmental law. It provides a range of services regarding environmental publications and statistics in the United States.

Fish and Wildlife Service (USFWS)
1849 C Street NW, Room 3012
Washington, DC 20240
(202) 208-7535
E-mail: Web_Reply@mail.fws.gov
Website: http://www.fws.gov/

The mission of the Fish and Wildlife Service is "to conserve, protect and enhance fish and wildlife and their habitats for the benefit of the American people." A division of the Department of the Interior, the USFWS has seven regional offices: Pacific, Southwest, Great Lakes-Big Rivers, Southeast, Northeast, Mountain Prairie, and Alaska.

United States Forest Service (USFS)
P.O. Box 96090
Washington, DC 20090-6090
(202) 205-1760
Website: http://www.fs.fed.us

The Forest Service is responsible for managing the national forests. Its mission is "to care for land and serve people," and it is bound by law "to achieve quality land management under the sustainable multiuse management concept to meet people's diverse needs." It has a number of regional offices throughout the United States.

Publications: The Forest Service publishes a variety of reports, guides, maps, and videos on the national forests and forestry issues. It also makes a number of publications available online.

National Professional Organizations
There are a number of national and international professional organizations to which those working in environmental ethics may wish to belong.

American Philosophical Association (APA)
c/o Janet Sample
University of Delaware
Newark, DE 19716
(302) 831-4657
E-mail: jsample@brahms.udel.edu
Website: http://www.oxy.edu/apa/apa.html

The main professional organization for U.S. philosophers. It holds frequent conferences, at which the International Society for Environmental Ethics may also hold meetings. The APA also supports an electronic bulletin board, accessible from the website listed above.

Association for the Study of Literature and the Environment
University of Delaware, Parallel Programs
333 Shipley Street
Wilmington, DE 19801
(302) 573-5463
E-mail: teague@strauss.udel.edu
Website: http://faraday.clas.virginia.edu/~djp2n/asle.html

Founded in 1992, this organization explores ideas and information about literature, which considers the relationship between human beings and the natural world. It also promotes new nature writing and interdisciplinary environmental research. It produces a handbook for graduate study in the area, a collection of essays on nature writing, and a bibliography of scholarship.

Association of Forest Service Employees for Environmental Ethics (AFSEE)
P.O. Box 11615
Eugene, OR 97440
(541) 484-2692
Website: http://www.afseee.org

This nonprofit organization seeks to forge a socially responsible value system for the Forest Service based on a land ethic that ensures ecologically and economically sustainable resource management. Membership is open to concerned citizens as well as those working in the Forest Service.

Publications: *Inner Voice: Forest Service Employees Speaking as Concerned Citizens* (bimonthly journal).

Association of Practical and Professional Ethics
410 North Park Avenue
Bloomington, IN 47405
(812) 855-6450
Website: http://ezinfo.ucs.indiana.edu/~appe/home.html

This association is committed to encouraging high-quality scholarship and teaching in practical and professional ethics (including environmental ethics) and supports development and research. It is a particularly important organization for those with a professional interest in ethics, including environmental ethics.

National Association of Environmental Professionals (NAEP)
5165 MacArthur Boulevard NW
Washington, DC 20016
(202) 966-1500
Fax: (202) 966-1977
Website: http://enfo.com/NAEP/

The NAEP is a multidisciplinary association for environmental professionals, providing a network of contacts and information on environmental planning, research, and management. Members must be environmental professionals and commit themselves to acceptance of a code of ethics and standards of practice put forward by the organization.

International Professional Organizations

International Society of Environmental Ethics
c/o Laura Westra, Secretary
Department of Philosophy
University of Windsor
Windsor, Ontario
N9B 3P4
Canada
Website: http://cep.unt.edu/ISEE.html

This is the major international organization for those with a professional or personal interest in environmental ethics. It exists to encourage research and education in environmental philosophy, including the philosophy of nature. It aims to promote human understanding of, respect for, and conservation of the natural world. Members come from a variety of backgrounds and from all over the world. The ISEE website gives further details of the society and provides access to a searchable bibliography of environmental ethics literature.

Publications: *International Society for Environmental Ethics Newsletter* (quarterly).

Society for Philosophy and Geography
c/o Jonathan M. Smith
Department of Geography
Texas A&M University
College Station, TX 77843-3147
E-mail: J0S7507@tamvm1.tamu.edu
Website: http://www.cep.unt.edu/geosoc.html

This international society aims to bring together the work of professional geographers and philosophers in areas where their work overlaps. One of these areas is environmental ethics, and the first issue of the society journal, *Philosophy and Geography* (1996), was devoted to environmental ethics.

Publications: *Philosophy and Geography* (annual journal).

CONTEMPORARY ETHICAL ISSUES

Chapter 8:
Selected Print
Resources

The field of environmental ethics has expanded rapidly, generating a vast number of publications. This chapter provides a selective annotated bibliography, highlighting some of the most important journals, books, and reports in the field, which should assist readers in selecting those that will be most useful to them.

Reference Works

Brown, Lester, et al., eds. *State of the World: A Worldwatch Institute Report on Progress toward a Sustainable Society.* New York: W. W. Norton.

This annual publication provides an overview of the state of the global environment. It is a useful reference work for all environmental fields, including environmental ethics.

Callicott, J. Baird. 1994. *Earth's Insights: A Multicultural Survey of Ecological Ethics from the Mediterranean Basin to the Australian Outback.* Berkeley: University of California Press. 285 pp. ISBN 0-520-08559-0.

A useful work for those interested in exploring different international approaches to environmental ethics. Callicott surveys a wide variety of understandings of environmental ethics, from Western European to Asian, South American, African, and Australasian. He also examines "postmodern" environmental ethics and some examples of environmental ethics in practice.

Childres, J., and J. MacQuarrie, eds. 1986. *A New Dictionary of Christian Ethics* (published in the United States as *The New Westminster Dictionary of Christian Ethics*). London: Westminster Press and SCM Press. ISBN 0-334-02205-3.

This dictionary contains entries, largely from a Christian perspective, on animals (by Andrew Linzey) and on environmental ethics (by Terence Anderson).

Clarke, Paul Barry, and Andrew Linzey, eds. 1996. *Dictionary of Ethics, Theology and Society*. New York: Routledge. 925 pp. ISBN 0-415-06212-8.

A sizable (and rather expensive) interdisciplinary reference work with entries on a wide variety of issues relevant to environmental ethics. The entry on environment (pp. 289–294) is written by J. Ronald Engel and R. J. S. Montagne. The book also contains entries on zoos and animal rights.

Cunningham, W., et al., eds. 1994. *Environmental Encyclopaedia*. Detroit: Gale Research International. ISBN 0-8103-4986-8.

A substantial encyclopedia on environmental topics, this book contains a number of useful biographies as well as entries on particular environmental institutions, laws, issues, and problems. Entries, which vary in length, are all fairly detailed and recommend articles and books for further reading.

Environment Abstracts. New York: R. R. Bowker. Monthly with annual cumulated editions.

A good place to find academic and technical papers on the whole range of environmental issues.

Environmental Philosophy: An Introductory Survey. 33 pp.
Environmental Philosophy: A Bibliography. 78 pp.
Ethics for Environmentalists. 12 pp.

These three booklets were jointly published by the Centre for Philosophy and Public Affairs, University of St. Andrews, U.K., and the U.K. Nature Conservancy Council in 1990. They can be obtained from the Director, Centre for Philosophy and Public Affairs, Department of Moral Philosophy, University of St. Andrews, St. Andrews, Fife, KY16 9AL, Scotland, UK. Although they are becoming a little outdated, they provide useful introductory and reference material.

Gilpin, Alan. 1996. *Dictionary of Environment and Sustainable Development.* Chichester, West Sussex, UK: John Wiley & Sons. 247 pp. ISBN 0-471-96220-1.

A useful introduction to key terms relating to the environment and sustainable development.

Mautner, Thomas, ed. 1996. *A Dictionary of Philosophy.* Oxford, UK: Basil Blackwell. 482 pp. ISBN 0-631-18459-7.

This dictionary of philosophy makes no explicit mention of environmental ethics. However, it is helpful for defining key philosophical terms and for brief introductions to general ethical concepts.

McInerney, Peter. 1992. *Introduction to Philosophy.* New York: Harper-Collins College Outline series. 242 pp. ISBN 0-06-467124-0.

An introductory philosophy text designed for first-year college students. It contains an introduction to major philosophers, a glossary of key terms, and a description of key moral theories.

Singer, Peter, ed. 1993. *A Companion to Ethics.* Corrected version. Oxford, UK: Blackwell. ISBN 0-631-18785-5.

A useful reference work, covering a range of topics in ethics. It contains helpful entries on animals by Lori Gruen and on environmental ethics by Robert Elliott.

————. 1994. *Ethics.* Oxford, UK: Oxford University Press. 415 pp. ISBN 0-19-289245-2.

A collection of short extracts on the nature and history of ethics, with sections on different approaches to ethics such as natural law and utilitarian approaches. Also contains several extracts of relevance to environmental ethics, especially Mary Midgley's article, "Duties Concerning Islands."

Thompson, Mel. 1994. *Teach Yourself Ethics.* London: Hodder and Stoughton Educational. 221 pp. ISBN 0-340-61101-4.

An elementary introduction to ethics, different theories of ethics, and themes in ethics. Section 9, "A Global Perspective," includes introductory material on animal rights and environmental ethics.

World Resources. New York: Oxford University Press, yearly publication. ISBN 0-19-521161-8.

This series was produced by collaboration among the World Resources Institute, the United Nations Development Program, the United Nations Environment Program, and the World Bank. It provides up-to-date reports on basic conditions, trends, and key issues in the global environment.

Books

Armstrong, S., and R. Botzler, eds. 1993. *Environmental Ethics: Divergence and Convergence*. New York: McGraw-Hill. 570 pp. ISBN 0-07-002608-4.

This textbook contains a number of extracts from a variety of key works in environmental ethics. It is divided into sections on different perspectives on environmental ethics (such as individualism, ecocentrism, Judeo-Christian, ecofeminism) within which a number of short extracts are gathered, introduced by the editors. The book is a very helpful teaching tool but is more generally useful as an introduction to environmental ethics and as a pointer to the key areas of the field and material published in it.

Attfield, Robin. 1991. *The Ethics of Environmental Concern*. Revision of 1983 ed. Athens: University of Georgia Press. 249 pp. ISBN 0-8203-1344-0.

Attfield's book explores Judeo-Christian ideas of stewardship and domination, as well as considering humans' obligations to future generations of humans and to the nonhuman natural world. The revised edition also contains a bibliographic essay and a new introduction.

Bookchin, Murray. 1982. *The Ecology of Freedom: The Emergence and Dissolution of Hierarchy*. Palo Alto, CA: Cheshire Books. 385 pp. ISBN 0-917352-10-6.

This book, by the "founder" of social ecology, explains his ideas about a non-hierarchical society based on ecological principles living in harmony with nature. It develops Bookchin's portrayal of social ecology and has important implications for environmental ethics.

Brennan, Andrew. 1988. *Thinking about Nature: An Investigation into Nature, Value and Ecology*. Athens: University of Georgia Press. 235 pp. ISBN 0-415-00303-2.

This book includes an important and helpful critical investigation of the way "scientific" ecology is used in some forms of "ecological" philosophy and ethics, especially deep ecology. The author proposes his own approach to environmental philosophy, which he calls ecological humanism.

Carson, Rachel. 1962. *Silent Spring*. New York: Penguin. 317 pp. ISBN 0-14-013891-9.

This environmental classic is often thought to be the book that began the environmental movement. Carson argues that the toxic effects of insecticide and other agricultural chemical residues on wildlife and human health could be catastrophic for human beings and other forms of life. She argues that humans need to adopt other, nontoxic methods of controlling crop pests and diseases.

Cooper, David, and Joy Palmer, eds. 1992. *The Environment in Question: Ethics and Global Issues*. New York: Routledge. 256 pp. ISBN 0-415-04968-7.

A useful collection of essays on a wide variety of issues in environmental ethics, including poverty, animals, tourism, radioactive waste, rain forests, pollution, and sustainability, as well as papers reflecting on the nature of environmental ethics itself. David Cooper's paper, "The Idea of Environment," is helpful for clarification of terms, and Rosemary Stephenson's paper, "Thinking, Believing, Persuading: Some Issues for Environmental Activists," is useful in thinking about what people believe about environmental issues and why.

————. 1995. *Just Environments: Intergenerational, International and Interspecies Issues*. New York: Routledge. 199 pp. ISBN 0-415-10336-3

A second useful collection of essays containing articles on issues of environment, development, and future generations. It includes essays by a range of contributors, including the editors themselves.

Devall, W., and D. Sessions. 1985. *Deep Ecology: Living as If Nature Mattered*. Layton, UT: Peregrine Smith. 267 pp. ISBN 0-87905-247-3.

An introduction to the basic ideas of deep ecology by two of its leading exponents. The authors discuss the "modern worldview" and that of "reformist environmentalism" before proposing their own, alternative way of viewing the world from a deep ecological perspective.

Elliott, R., and A. Gare, eds. 1983. *Environmental Philosophy*. Queensland, Australia: Open University Press. 303 pp. ISBN 0-335-10407-X.

An early collection of papers that was for many years the classic anthology in environmental ethics. It includes papers on a variety of topics, including the nature of an environmental ethic, environmental policy and human welfare, and attitudes toward the natural environment in different cultural traditions.

Engel, J. R., and J. G. Engel, eds. 1990. *Ethics of Environment and Development: Global Challenge and International Response*. London: Belhaven. 264 pp. ISBN 1-85293-084-5.

This collection sprang from the Ethics Working Group of the International Union for the Conservation of Nature and Natural Resources (IUCN). It brings together a number of papers on ethical questions in sustainable development from a range of different cultural perspectives, and includes contributors such as Arne Naess and Holmes Rolston. The introductory essay by J. Ronald Engel, "The Ethics of Sustainable Development," is particularly interesting.

Hargrove, Eugene, ed. 1992. *The Animal Rights/Environmental Ethics Debate: The Environmental Perspective.* New York: State University of New York Press. 273 pp. ISBN 0-7914-0934-1.

This anthology contains many of the key articles in the animal rights–environmental ethics debate, including J. Baird Callicott's important paper "Animal Liberation: A Triangular Affair." It is a convenient way of accessing papers spread through a range of journals and also has a helpful historical introduction.

Johnson, Lawrence. 1991. *A Morally Deep World: An Essay on Moral Significance and Environmental Ethics.* New York: Cambridge University Press. 301 pp. ISBN 0-521-39310-8.

This is a systematic work in environmental ethics. Johnson argues that individual living organisms and ecological collectives such as ecosystems and species are entities in their own right and should be taken into account when making moral decisions.

Leopold, Aldo. 1947. *A Sand County Almanac and Sketches Here and There.* New York: Oxford University Press. 226 pp. ISBN 0-19-500777-8.

A collection of classic essays by the writer and forester Aldo Leopold, first published after his death in 1947. The book includes his essays "Conservation Esthetic," "Wilderness," and most importantly his influential "The Land Ethic."

List, Peter, ed. 1993. *Radical Environmentalism: Philosophy and Tactics.* Belmont, CA: Wadsworth. 276 pp. ISBN 0-534-17790-5.

List's collection of extracts focuses on the philosophies and activities of a range of radical environmental groups. In the first part of the book, he includes extracts from key writers in deep ecology, ecofeminism, social ecology, and bioregionalism; in the second part of the book, he includes work by environmental activists in Greenpeace and Earth First! and some philosophical responses to these activities.

Lovelock, James. 1979. *Gaia: A New Look at Life on Earth.* New York: Oxford University Press. 157 pp. ISBN 0-19-286030-5.

This is the famous book in which Lovelock puts forward his controversial hypothesis that the earth behaves like a living organism, keeping the earth hospitable to life. The book includes some suggestions interesting for environmental ethics.

Mathews, Freya. 1991. *The Ecological Self.* London: Routledge. 192 pp. ISBN 0-415-05252-1.

Mathews explores the metaphysical foundations of environmental ethics, in particular the idea that humans are "one" with nature. She focuses on ideas of interconnectedness. This is a useful book for those interested in developing their understanding of key ideas in deep ecology.

Merchant, Carolyn. 1992. *Radical Ecology: The Search for a Livable World.* New York: Routledge. 276 pp. ISBN 0-415-90650-4.

An introductory book on different approaches to radical ecology. It includes sections on deep ecology, ecofeminism, social ecology, and green politics.

Midgley, Mary. 1983. *Animals and Why They Matter.* London: Penguin. 153 pp. ISBN 0-820-30704-1.

This book provides a clear introduction to the history and philosophies behind Western attitudes toward and relationships with animals. Midgley explores a number of influential philosophical approaches to animals and considers new ways humans might relate to animals.

Nash, Roderick. 1989. *The Rights of Nature: A History of Environmental Ethics.* Madison: University of Wisconsin Press. 290 pp. ISBN 0-299-11840-1.

The standard history of the development of environmental ethics (also environmental philosophy more generally) until the late 1980s. This is a useful and informative text, though it is sometimes criticized for discussing environmental ethics largely in terms of rights.

O'Neill, John. 1993. *Ecology, Policy and Politics: Human Well-Being and the Natural World.* New York: Routledge. 227 pp. ISBN 0-415-07300-6.

One of the books in the useful Routledge Environmental Philosophy series. O'Neill adopts a "virtues" approach to environmental ethics and applies it to environmental policy making. The book is particularly useful for its clear exposition, in Chapter 2, of the different ways in which the term "intrinsic value" may be used in environmental ethics.

Pojman, Louis. 1994. *Environmental Ethics: Readings in Theory and Application.* Sudbury, MA: Jones and Bartlett. 503 pp. ISBN 0-86720-951-8.

A collection of readings and study questions, this is a textbook designed for use in classroom teaching.

Regan, Tom. 1984. *The Case for Animal Rights.* London: Routledge and Kegan Paul. 425 pp. ISBN 0-7102-0150-8.

This is a classic work in the animal liberation tradition. Regan argues systematically that like human beings, adult mammals have rights to life and that such rights should be protected. He discusses a number of additional issues raised by this position including hunting, vegetarianism, and the relationship between animal rights and environmental ethics.

Rolston, Holmes. 1986. *Philosophy Gone Wild.* New York: Prometheus Books. 269 pp. ISBN 0-87975-556-3.

A collection of key essays by an important writer in environmental ethics. His essays address such questions as whether there is an ecologic ethic, what we mean when we talk about values in nature, and what duties we have toward endangered species.

————. 1988. *Environmental Ethics: Duties to and Values in the Natural World.* Philadelphia: Temple University Press. 391 pp. ISBN 0-87722-501-X.

This is Holmes Rolston's major systematic work in environmental ethics. He examines a range of questions in environmental ethics, developing his own view on the nature of environmental value. He also includes sections on environmental policy and environment and business viewed from his position in environmental ethics.

————. 1994. *Conserving Natural Value.* New York: Columbia University Press. 259 pp. ISBN 0-231-07901-X.

A recent work by a leading writer in environmental ethics, this book introduces many of the ethical and philosophical issues at stake in biological conservation. It contains a number of interesting case studies.

de-Shalit, Avner. 1995. *Why Posterity Matters: Environmental Policies and Future Generations.* New York: Routledge. 161 pp. ISBN 0-415-10019-4.

This book is a systematic study of our obligations to future generations of humans, arguing that we have obligations to them as a matter of justice. This idea has implications for environmental policy making and for environmental ethics, which are considered in the book.

Singer, Peter. 1975. *Animal Liberation.* New York: Thorsons, Harper-Collins. 320 pp. ISBN 0-7225-2415-3.

This classic work is often credited with founding the modern animal liberation tradition. Singer argues that because of their capacities to feel pleasure and pain, animals have interests that should be taken into account when we make ethical decisions. On these grounds he argues against a variety of practices involving animals, including the use of animals in scientific research and factory farms. The reprinted edition has a new introduction.

Stone, Christopher. 1987. *Earth and Other Ethics: The Case for Moral Pluralism.* New York: Harper and Row. 280 pp. ISBN 0-06-091486-6.

In this more recent book, Stone argues that no single traditional ethical system can successfully deal with the difficult issues raised by environmental ethics, in particular the question of the "rights" of natural objects. He argues that a more flexible, pluralistic approach to ethics is needed. His controversial arguments here have been important in the developing discussion over moral pluralism in environmental ethics.

————. 1988. *Should Trees Have Standing? Toward Natural Rights for Legal Objects.* Palo Alto, CA: Tioga. 102 pp. ISBN 0-935382-69-0.

This is an updated version, with a new preface and introduction, of an article that Christopher Stone, a lawyer, published in the *Southern Californian Law Review* in 1972. The article was written with reference to a particular case about wilderness development then in the courts, to argue that natural objects should have legal standing in their own right, just as (for instance) corporations do. This article has subsequently been influential in environmental ethics as well as in the development of environmental law.

Sylvan, R., and D. Bennett. 1994. *The Greening of Ethics.* Tuscon: University of Arizona Press. 269 pp. ISBN 0-874267-04-9 (British edition).

The authors examine several different approaches to environmental ethics before proposing their own alternative, Deep Green Theory, which is discussed from an Australian perspective. This is one of the few easily accessible texts written by the Australian environmental philosopher Richard Sylvan, who died in 1996.

Taylor, Paul. 1986. *Respect for Nature: A Theory of Environmental Ethics.* Princeton, NJ: Princeton University Press. 329 pp. ISBN 0-691-02250-X.

This is a classic work in environmental ethics. Taylor proposes a systematic, thoroughgoing approach to environmental ethics based on the principle of respect for nature. He argues that all living organisms have their own inherent worth, and that all have it equally. Where this leads to situations of conflict, Taylor proposes a series of priority principles for conflict resolution. Some readers may find Taylor's arguments rather dense, but his book is an important contribution to environmental ethics.

Tucker, Mary Evelyn, and John Grim, eds. 1994. *Worldviews and Ecology: Religion, Philosophy and the Environment.* New York: Orbis Books. 246 pp. ISBN 0-88344-967-6.

A collection of papers exploring the approaches of different philosophical and religious worldviews to environmental issues and sustainability. It includes papers from the perspectives of major world religions, including Judaism, Islam, and Hinduism, as well as a number of other modern perspectives on ecological questions.

Warren, Karen, ed. 1994. *Ecological Feminism*. New York: Routledge. 209 pp. ISBN 0-415-07298-0.

A collection of papers from a range of different approaches to ecofeminism. Many of the papers suggest ecofeminist approaches to and developments of environmental ethics.

Zimmerman, M., et al. 1993. *Environmental Philosophy: From Animal Rights to Radical Ecology*. Englewood Cliffs, NJ: Prentice-Hall. 437 pp. ISBN 0-13-666-959-X.

A collection of important papers in environmental philosophy helpfully brought together in a single volume. Areas covered include individualistic and holistic approaches to environmental ethics and deep ecology, social ecology, and ecofeminist approaches.

Selected Articles

Callicott, J. Baird. 1980. "Animal Liberation: A Triangular Affair." *Environmental Ethics* 2.

In this important article, Callicott argues that the animal liberation tradition and the environmental ethics traditions are not natural allies but based on fundamentally different ways of looking at the world. The article has been the subject of dispute and discussion in environmental ethics ever since.

Naess, Arne. 1973. "The Shallow and the Deep, Long-Range Ecology Movement: A Summary." *Inquiry* 16.

In this article, the expression *deep ecology* (opposed to shallow ecology) is used by Arne Naess for the first time. The article became the rallying point for the influential deep ecology movement and is therefore historically as well as philosophically important.

Regan, Tom. 1981. "The Nature and Possibility of an Environmental Ethic." *Environmental Ethics* 3.

This article is an early exploration of the difficulties involved in the construction of an environmental ethic. He considers the possible moral significance of nonliving natural objects and argues for an attitude of "admiring respect" toward the natural world.

Rolston, Holmes. 1975. "Is There an Ecological Ethic?" *Ethics* 85.

This article is another early and important exploration of the need for, and nature of, environmental ethics.

Russow, Lilly-Marlene. 1981. "Why Do Species Matter?" *Environmental Ethics* 3.

Russow challenges the widely accepted belief that the preservation of species is important. She questions common understandings of the term *species* and argues that we have moral obligations to individual members of species rather than to species as a whole.

Van De Veer, Donald. 1979. "Interspecific Justice." *Inquiry* 22.

Van De Veer explores ways to resolve interspecies conflicts. He proposes what he calls two-factor egalitarianism as a way to address such conflicts.

White, Lynn. 1967. "The Historic Roots of Our Ecologic Crisis." *Science* 155.

This article, which has been reprinted in a number of essay collections, triggered a still-running debate within Christianity about appropriate ethical attitudes to the environment. White, a historian, argued that the Judeo-Christian tradition carries a burden of responsibility for the environmental crisis due to its human-centeredness and its refusal to accept the idea that natural objects might be sacred. He does, however, suggest that some forms of Christianity, such as that followed by St. Francis of Assisi, might be more amenable to the development of environmental sensitivity.

Reports and Proceedings

Attfield, Robin, and Katharine Dell. 1989. *Values, Conflict and the Environment.* Report of the Environmental Ethics Working Party, Ian Ramsey Centre, St. Cross College, Oxford, UK.

A report produced by a multidisciplinary research group. It discusses attitudes and arguments frequently found in environmental conflicts, briefly examines how existing Western philosophies have responded to questions in environmental ethics, and proposes a method of making decisions in environmental ethics called "comprehensive weighting." It also contains a "minority report" outlining a different approach to environmental ethics.

Our Common Future. Report of the World Commission on Environment and Development, Oxford, UK: Oxford University Press, 1987. 400 pp. ISBN 0-19-282080-X.

Sometimes known as the Brundtland Report, this document was an important milestone in international environmental and developmental policy. It is a useful reference work for environmental ethics for this reason, but also for its discussions of needs, concerns, and priorities in environment and development, and for its widely used definition of sustainable development in Chapter 1.

Periodicals

The principal periodicals in the field of environmental ethics are listed below. The addresses given are subscription addresses. Periodicals in other fields (such as applied ethics or sociology) may also carry articles of relevance to environmental ethics, while wilderness preservation and other environmental organizations often produce their own high-quality periodicals.

Agriculture and Human Values
Department of Philosophy
University of Florida
330 Griffin-Floyd Hall
Gainesville, FL 32611

This is the journal produced by the Agriculture, Food and Human Values Society.

Between the Species
P.O. Box 8496
Landscape Station
Berkeley, CA 94707
(510) 526-5346

This journal is produced by the Schweitzer Center of the San Francisco Bay Institute/Congress of Cultures and publishes philosophical papers on the relationship between human beings and other species.

The Ecologist
RED Computing, Ltd.
The Outback
58-60 Kingston Road
New Malden
Surrey KT3 3LZ
UK

An international environmental journal that acts as a forum for social and environmental activists who seek to change development policies in the developing and the developed world.

Environmental Ethics
Center for Environmental Philosophy
P.O. Box 13496

University of North Texas
Denton, TX 76203-6496
(817) 565-2727
Fax: (817) 565-4448
E-mail: ee@unt.edu

This quarterly publication is the oldest and most important journal in the field, publishing articles, discussion papers, book reviews, and news relevant to environmental ethics.

Environmental Politics
Frank Cass & Co., Ltd.
Newbury House
900 Eastern Avenue
Newbury Park
London IG2 7HH
UK
44 (0)181 599 8866
Fax: 44 (0)181 599 0984
E-mail: jnlsubs@frankcass.com

This quarterly publication contains papers, research notes, review articles, and book reviews pertaining to both theoretical and applied areas in environmental politics. Many of the articles deal with issues in environmental ethics as they are relevant to politics.

Environmental Values
White Horse Press
1 Strond
Isle of Harris
Scotland
PA83 3UD

A U.K.-based quarterly publication bringing together contributions from philosophy, law, economics, and other disciplines to discussion of the foundations of environmental policy.

Ethics and the Environment
c/o JAI Press, Inc.
55 Old Post Road No. 2
P.O. Box 1678
Greenwich, CT 06836-1678

This quarterly journal, first published in 1996, is an interdisciplinary forum for theoretical and practical articles, discussions, reviews, and comments in environmental ethics and environmental philosophy.

Journal of Agricultural and Environmental Ethics
Kluwer Academic Publishers Group
Order Department
P.O. Box 322
3300 AH Dordrecht
The Netherlands
31-78-6392392
Fax: 31-78-6546474
E-mail: services@wkap.nl

A quarterly publication (first issue 1997), this journal publishes scientific articles and discussion papers on ethical issues confronting agriculture, food production, and the environment.

Radical Philosophy
Central Books (RP Subscriptions)
99 Wallis Road
London E9 1LN
UK

Although this journal is tagged "a journal of socialist and feminist philosophy," it also publishes a significant number of articles on environmental philosophy and environmental ethics, as well as reviews of books in the field.

Terra Nova: Nature and Culture
New Jersey Institute of Technology
University Heights
Newark, NJ 17102

A cross-disciplinary, refereed journal exploring a range of approaches to nature and culture.

Trumpeter
Lightstar
Box 5853 Stn. B
Victoria BC
Canada
V8R 6S8

A quarterly journal of ecophilosophy, publishing articles about a wide range of philosophical approaches to the natural world, including deep ecology.

Worldviews: Environment, Culture Religion
White Horse Press
1 Strond
Isle of Harris
Scotland
PA83 3UD

A new interdisciplinary journal exploring the environmental understandings, perceptions, and practices of a wide range of cultures and religious traditions.

CONTEMPORARY ETHICAL ISSUES

Chapter 9: Selected Nonprint Resources

In the last two decades there has been an explosion in home video and computer technology. For this reason, there is a constantly increasing range of material available on environmental ethics. Only a small selection can be included here. For readers with access to electronic resources on the Internet, searches on relevant websites and membership in relevant forums will provide up-to-date material. This is true even of the video information, since many video companies have up-to-date online catalogs. In addition to the resources listed below, many environmental and animal welfare organizations such as Greenpeace and the Audubon Society produce their own videos on a variety of topics relevant to environmental ethics.

Videos and Films

All videos listed here are available in VHS format only.

Animal Rights

Length:	28 minutes
Date:	1996
Cost:	$49.95
Source:	Jones and Bartlett
	40 Tall Pine Drive
	Sudbury, MA 01776
	(800) 832-0034

One of an *Ethics in the 1990s* series produced by Joram Graf Haber for Cable Television Network, New Jersey. In this video, Professor Tom Regan, author of *The Case for Animal Rights*, is interviewed by Joram Graf Haber. Regan argues in favor of an animal rights view and deals with questions opposing this view.

Beauty and the Beasts

Length:	52 minutes
Date:	1995
Cost:	$149
Source:	Films for the Humanities and Sciences
	P.O. Box 2053
	Princeton, NJ 08543-2053
	(800) 257-5126
	Fax: (609) 275-3797

A program investigating human responses to the appearances of different kinds of animals and how this influences human treatment of them.

Energy and Morality

Length:	33 minutes
Date:	1981
Cost:	$275
Source:	Bullfrog Films
	Box 149
	Olney, PA 19547
	(610) 779-8226
	Fax: (610) 370-1978
	E-mail: bullfrog@igc.apc.org
	Website: http://www.bullfrogfilms.com/

This film relates energy use to different kinds of value systems. It focuses on the views of soft energy specialist Amory Lovins and the economist E. F. Schumacher, who suggest that human economies should adapt to the natural world rather than the other way around.

Environmental Ethics

Length:	28 minutes
Date:	1995

Cost: $49.95
Source: Jones and Bartlett
                  40 Tall Pine Drive
                  Sudbury, MA 01776
                  (800) 832-0034

Episode from *Ethics in the 1990s* series for Cable TV Network of New Jersey. Features an interview with the environmental ethicist Eric Katz, explaining what environmental ethics is and discussing some of the key issues in environmental ethics.

From the Heart of the World
Length: 85 minutes
Date: 1990
Cost: £99 (English-made film)
Source: BBC Videos for Education and Training
                  80 Wood Lane
                  London, W12 0TT
                  England
                  Fax: 44 (0)181 576 2916

Note: Only educational and training establishments can purchase this film.

A film about the ancient Kogi people of Colombia, who have previously remained hidden but have spoken to the cameras in an attempt to try to save the world from environmental catastrophe. An interesting "alternative" understanding of humans' ethical obligations to the natural world.

Gaia Theory with James Lovelock
Length: 52 minutes
Date: 1995
Cost: £15 + 15% postage (English-produced film)
Source: Green Books Ltd.
                  Foxhole
                  Dartington, Totnes
                  Devon TQ9 6EB
                  UK
                  (0)1803 863260
                  Fax: (0)1803 863 843

James Lovelock, inventor of the Gaia Theory, explains his ideas for those with little scientific knowledge. The video is useful for study groups and could form the basis of a discussion on environmental ethics.

Icon Earth
Length: 50 minutes
Date: 1995

Cost: £99 (English-made film)
Source: BBC Videos for Education and Training
           80 Wood Lane
           London, W12 0TT
           UK
           Fax: 44 (0)181 576 2916

Note: Only educational and training establishments can purchase this film.

A BBC documentary that explores how technological advances have changed our perceptions and understanding of the planet.

In Defense of Animals: A Portrait of Peter Singer

Length: 28 minutes
Date: 1989
Cost: $250
Source: Bullfrog Films
           Box 149
           Olney, PA 19547
           (610) 779-8226
           Fax: (610) 370-1978
           E-mail: bullfrog@igc.apc.org
           Website: http://www.bullfrogfilms.com/

A film featuring an interview with the philosopher Peter Singer, author of the classic work *Animal Liberation*. Singer presents the argument that he believes underpins the animal liberation movement. The video is accompanied by a study guide.

Introduction to Ecological Economics

Length: 45 minutes
Date: 1991
Cost: $25
Source: Griesinger Films
           7300 Old Mill Road
           Gates Mills, OH 44040
           Tel/Fax: (216) 423-1601

Introduces a variety of areas of ecological economics, including Natural Capital, the Index for Sustainable Economic Welfare, and Empiricism and Values. It argues that current systems of economic accounting fail to appropriately measure social well-being or ecological health, and it discusses related ethical questions.

John Muir's High Sierras

Length: 27 minutes (school version 21.5 minutes)
Date: 1973

Cost: $310 (school version $290)
Source: Churchill Films
               662 North Robertson Boulevard
               Los Angeles, CA 90069

This film covers the route taken by John Muir on some of his hikes in Yosemite. Extracts from his diaries are read over the pictures.

The Mighty River

Length: 24 minutes
Date: 1993
Cost: Institutions $89; individuals $35
Source: The Video Project
               200 Estates Drive
               Ben Lomond, CA 95005
               (408) 336-0160
               Fax: (408) 336-2168

This is a highly acclaimed animated short, nominated for an Academy Award and narrated by actor Donald Sutherland. It focuses on the St. Lawrence River and the negative impact humans have had on the river over the years, and could be used to stimulate discussion on human ethical responsibilities to the natural world.

Oxford Centre for Environment, Ethics and Society: Debates, 1996

Length: 90 minutes each
Date: 1996
Cost: £17 (including postage)
Source: Oxford Centre for Environment, Ethics and Society
               Mansfield College
               Oxford
               OX1 3TF
               UK
               Tel and Fax: 44(0)1865 270 886
               E-mail: ocees@mansfield.ox.ac.uk
               Website: http://users.ox.ac.uk/~ocees

Four videos of head-to-head debates from 1996. The topics and debaters are "Indigenous Peoples as Conservationists," Aroha Te Pareake Mead and Kent Redford; "Tourism, Travel and the Environment," Geoffrey Lipman and Paul Gonsalves; "Voluntary Simplicity," Duane Elgin and Juliet Schor; "Global Environmental Concern: At Whose Expense?," Sunita Narain and Andrew Steer.

Pelts: Politics of the Fur Trade

Length: 56 minutes
Date: 1990

Cost: $250
Source: Bullfrog Films
            Box 149
            Olney, PA 19547
            (610) 779-8226
            Fax: (610) 370-1978
            E-mail: bullfrog@igc.apc.org
            Website: http://www.bullfrogfilms.com/

Produced by the National Film Board of Canada, this film explores a variety of political and ethical perspectives on the Canadian fur trade. Representatives of different sides in the dispute are interviewed and the tactics of both sides investigated.

Prophets and Loss
Length: 49 minutes
Date: 1991
Cost: Institutions $79; Individuals $35
Source: The Video Project
            200 Estates Drive
            Ben Lomond, CA 95005
            (408) 336-0160
            Fax: (408) 336-2168

Featuring interviews with key thinkers, including Carl Sagan and Paul Ehrlich, this video explores the links between the global problems of poverty and environmental destruction and spiritual and philosophical values.

Race to Save the Planet
Length: 15 minutes each
Date: 1989
Cost: $89.95 each ($599 series)
Source: Films for the Humanities and Sciences
            P.O. Box 2053
            Princeton, NJ 08543-2053
            (800) 257-5126
            Fax: (609) 275-3797

A series of seven short videos designed as teaching modules, with teacher's guide. The programs use a case-study approach and cover the following issues: Saving the Land; Saving the Atmosphere; Saving the Water; Saving the Diversity of Life; Saving the Planet; Environmental Detectives; Marine Detectives.

Risky Business
Length: 24 minutes
Date: 1996

Cost: $195
Source: Bullfrog Films
Box 149
Olney, PA 19547
(610) 779-8226
Fax: (610) 370-1978
E-mail: bullfrog@igc.apc.org
Website: http://www.bullfrogfilms.com/

A short 1996 film introducing the ethical issues raised by the growth in biotechnology.

Where on Earth Are We Going?
Length: 25 minutes each
Date: 1990
Cost: £65 per episode; £210 series (British-made film)
Source: BBC Videos for Education and Training
80 Wood Lane
London, W12 0TT
UK
Fax: 44 (0)181 576 2916

Note: Only educational and training establishments can purchase these films.

A series of six programs: Energy and Pollution, Food and Agriculture, Green Society, Industry and Work, International Perspective—One World or No World, and Getting There—Body Politic, Immortal Soul. Presented by charismatic U.K. environmental campaigner Jonathan Porritt, this series has sold in more than 50 countries.

The Wilderness Idea—John Muir, Gifford Pinchot and the First Great Battle for Wilderness
Length: 58 minutes
Date: 1989
Cost: $34.95
Source: Direct Cinema, Ltd.
P.O. Box 10003
Santa Monica, CA 90410
(310) 396-4774

A film about the dispute, in the first decade of this century, over the damming of Hetch Hetchy valley to provide a water supply for San Francisco. A useful case study of a contest over development and wilderness.

Wilderness: The Last Stand
Length: 53 minutes
Date: 1993

Cost: Institutions $95; Individuals $45
Source: The Video Project
             200 Estates Drive
             Ben Lomond, CA 95005
             (408) 336-0160
             Fax: (408) 336-2168

An insightful investigation into the continued felling of virgin forest in the United States, including interviews with individuals on all sides in the debate. It is narrated by actress Susan Sarandon and is accompanied by a study guide.

CD-ROMs and Software

There are few resources specifically in environmental ethics available in these formats. However, some CD-ROM and other software resources raise ethical issues in an environmental context and provide case-study material around which ethical debate may be built.

Earth Aware

Format: Floppy disk for Windows 3.1, Windows 95, or Macintosh
Cost: Schools $69.95; individuals $34.95
Produced by: Enviroaccount Software
Source: Enviroaccount Software
               605 Sunset Court
               Davis, CA 95616
               Tel/Fax: (800) 554-0317
               E-mail: dwiotter@dcn.davis.ca.us
               Website: http://wheel.dcn.davis.ca.us/go/earthaware

This software allows individuals and groups to carry out a comprehensive personal environmental impact assessment (which gives a score, a rating on a scale from Eco-Titan to Eco-Tyrannosaurus Rex) and to find out their personal CO_2 total in kg/yr.

Earth Explorer

Format: CD-ROM
Cost: $49.95
Produced by: Apple Computer and Enteractive, Inc.
Source: The Video Project
               5332 College Avenue, Suite 101
               Oakland CA, 94618
               (510) 655-9050
               Fax: (510) 655-9115
Note: Specify Mac or PC version.

A CD-ROM program designed for children ten years old and upward to adults. It introduces 21 environmental topics, allowing for research and providing interactive information and educational games. It also has pro and con arguments on particular environmental issues, useful for debates in environmental ethics. It includes over 400 illustrated articles and sections concerning environmental data and maps.

Encyclopedia of U.S. Endangered Species

Format:	CD-ROM
Cost:	$49.95
Produced by:	ZCM
Source:	The Video Project
	5332 College Avenue, Suite 101
	Oakland, CA 94618
	(510) 655-9050
	Fax: (510) 655-9115
Note:	Specify Mac or PC version.

A reference guide to more than 700 listed endangered or threatened species in the United States. It includes information about the species (with text, location maps, legal status, sounds, and photos), bibliography, glossary, questions, and quizzes.

Environmental Abstracts

Format:	CD-ROM
Cost:	$1,360
Produced by:	Congressional Information Service, Inc.
Source:	Congressional Information Service, Inc.
	4520 East-West Highway
	Bethesda, MD 20814-3389
	(800) 638-8380 (from outside U.S.: +1-301-654-1550)
	Fax: (301) 654-1550
	E-mail: info@cispubs.com
Note:	Also available in print, on microfiche, and on magnetic tape for networked systems.

The Environmental Abstracts service provides abstracts for users (including full bibliographic information) from journal articles, conference papers and proceedings, and other sources. About 1,400 new items are added monthly, and material is divided into 21 major subject areas, including environmental education, environmental policy, population, and a range of other areas. Environmental ethics articles are also abstracted (although they do not fall into a separate category). On CD-ROM, there is a browse index, a search mode for a variety of different types of search, and a form search option.

SimEarth and other Sim products

Format: CD-ROM
Produced by: Maxis Ltd. 1995, from Turner Interactive
Source: All regular CD-ROM distributors
Note: Available for Macintosh, IBM/DOS, or Windows.

The family of Sim programs—including SimEarth, SimIsle, SimFarm, and SimCity—are CD-ROM simulation programs allowing the construction of planets, islands, farms, or cities. All have ecological themes. SimEarth is the best known of the programs; it allows for the design and nurture of planets from creation through evolution, inspired by James Lovelock's model of the earth as a living organism. SimIsle allows for the creation of island rain forest ecosystems, which have to be defended against miners, poachers, tourists, and polluters. SimFarm provides (among a range of options) the opportunity to run a simulated organic farm, while SimCity provides for the construction and maintenance of sustainable cities.

Worldwide Fund for Nature (WWF) Data Bulletins

Format: Floppy disk for PC or Macintosh
Cost: £19.99 for each Data Bulletin (British-produced)
            and £6.99 for tutor's guides
Source: WF-UK
            Education and Awareness
            Panda House, Weyside Park
            Godalming
            Surrey GU7 1XR
            UK
            44 (0)1483 426444
            Fax: 44 (0)1483 426409

The Worldwide Fund for Nature (WWF), known in the United States as the World Wildlife Fund, has established a data bulletin series on floppy disks. Data bulletins combine text, graphics, and data on specific environmental issues and can be exported into everyday software and networked as well as printed from disk. Teacher's guidance notes are also available. Currently, WWF is marketing a number of data bulletins. One is a double bulletin based on the U.K. controversy surrounding the dumping of an oil storage buoy, Brent Spar, at sea in 1995. It contains material from the protagonists, Shell and Greenpeace, and raises issues about the role of the media and pressure groups and ethical principles of care for the marine environment. Another, Tioxide Europe, examines the environmental choices that must be made when making decisions about the impact of industry on the natural environment. A third data bulletin, called New Economics, may be of particular interest to those working in environmental ethics, as it examines questions about the challenges environmental philosophy poses to economics.

Online Databases, Forums, and Websites

With the growth of the Internet, the usefulness of electronic communications in environmental ethics has grown significantly. In particular, E-mail discussion lists and environmental ethics web pages can provide a wealth of information on—and interaction about—environmental ethics. To participate in E-mail discussion, you must electronically join the relevant discussion group; you will then receive all mailings sent to the group and may mail to the whole group yourself. Web pages do not allow this direct interaction but do provide a significant amount of useful information and give you the facility of linking to other web pages in related areas.

E-Mail Forums

ecotheol

This discussion list is dedicated to the relationship of theology and religion to environmental issues and frequently includes discussion on environmental ethics from different theological and religious perspectives. To subscribe, send the message *join ecotheol your first name your last name* to mailbase@mailbase.ac.uk.

enviroethics

This international E-mail discussion group is entirely dedicated to the discussion of environmental ethics. It is also used to publicize relevant jobs, conferences, and other information. To subscribe, send the message *join enviroethics your first name your last name* to mailbase@mailbase.ac.uk.

geo-ethics

Geo-ethics is an international e-list that provides a forum for discussion and information regarding ethics and geography. Discussion can be wide-ranging (including development ethics and professional ethics for geographers), but list members also frequently discuss environmental ethics. To subscribe, send the message *SUBSCRIBE GEO-ETHICS your e-mail address* to MAJORDOMO@ATLAS.SOCSCI.UMN.EDU.

Internet Resources: World Wide Websites

Many useful websites have been created by national and international academic departments and environmental organizations. These web addresses have been listed alongside the relevant organizations in Chapter 7. However, some of the key web addresses and useful sites not listed in that chapter are included here.

EcoNet

http://www.econet.apc.org

EcoNet is a net service that connects individuals and organizations working in the field of environmental protection and sustainability. It supports a variety of web pages; it also contains a directory of environmental organizations, a resource center, a news section, and a section of featured items. It has a search facility for a range of topics.

EnviroLink

http://www.envirolink.org

EnviroLink is an international online environmental resource that links environmental organizations and volunteers with those who visit its site. It claims to be the largest online environmental resource in the world.

Environmental Protection Agency

http://www.epa.gov

This is the homepage of the U.S. Environmental Protection Agency.

International Society for Environmental Ethics (ISEE)

http://www.cep.unt.edu/ISEE.html

This website provides a database of members and officers of the ISEE and the ISEE bibliography compiled by Holmes Rolston. The bibliography is searchable by topic and author. Back copies of the ISEE newsletter are also accessible from this website.

University of North Texas, Center for Environmental Philosophy

http://www.cep.unt.edu

This site has general material on environmental ethics, including news in the field, graduate and professional study programs, and new books, as well as information about courses and staff in environmental philosophy at the University of North Texas. It provides helpful links to other sites of interest, including related organizations, journals, and the website and bibliography of the International Society for Environmental Ethics.

CONTEMPORARY ETHICAL ISSUES

Glossary

Acid Rain: Rain that contains dissolved pollutants, particularly sulfuric and nitric acid. These pollutants primarily come from coal-fired power stations and from vehicle exhausts. Acid rain can damage forests and kill living organisms in inland waters. It is more properly known as acid precipitation (since fog and snow can have the same effects).

Anarchist: Someone who believes that the ideal society would be one in which formalized systems of government and the institutions of the state are abolished.

Animal Liberation Movement: An umbrella term for the political movement that argues that animals are oppressed and unjustly treated in present societies. Individuals from a spectrum of different philosophical perspectives may associate themselves with this movement: some animal welfare campaigners, some who argue that animals have rights, and more radical direct-action groups.

Anthropocentric: Literally, this term means "human-centered." In environmental ethics, it is used to describe ethical positions where only humans are considered to be of direct moral concern. From such a perspective, the environment is only important because it contributes to human well-being.

Anthropogenic: Literally, this term means "of human genesis" or "originating from humans." It is important to avoid confusing this term with anthropocentric (above). Though all approaches to ethics are *anthropogenic* (in that humans create them), they need not be *anthropocentric* (concerned only with human well-being).

Biodiversity: An abbreviation for *biological diversity*—the existence of a wide variety of kinds of life on earth. It may be used to refer to the diversity of individual organisms, to the diversity of genetic material, or to the diversity of kinds of species in existence. In popular use, it usually refers to the latter.

Biosphere: The part of the earth's surface including land, water, and atmosphere that is inhabited by living things.

Chlorofluorocarbons (CFCs): A family of artificial chemicals first developed in the 1930s and used as refrigerants, propellants, and solvents. Once believed to be inert and harmless, CFCs are now implicated in the depletion of stratospheric ozone and global warming.

Climate Change: In its most general sense, climate change refers to long-term changes in weather patterns, usually on a global scale, whether or not these changes are of human origin. Most commonly, "climate change" is now used in a more particular sense to describe the changes in global climate that may be caused by increased emissions (by human beings) of gases that increase global warming.

Collectivist: In ethics, someone who values a group or collective (often understood as, in some sense, a whole) above an individual. In environmental ethics, the term usually describes someone who values the health of ecological systems and species more than the well-being of individual organisms.

Consequentialist: The philosophical belief that the *consequences* or results of actions (rather than their motivation or the character of the person who does them) are ethically significant. This view contrasts with a deontological or duty-based approach to ethics. The most popular form of consequentialism is *utilitarianism*.

Deep Ecology: A diverse philosophical movement originally associated with the work of the Norwegian philosopher Arne Naess. At first deep ecology was characterized by the conviction that humans should more fully recognize their interconnectedness with nature and should affirm that all living organisms are of equal intrinsic value. More recently there has been a considerable divergence of views within the deep ecology movement, although many deep ecologists still accept a foundational set of principles called the Deep Ecology Platform, drawn up in 1985.

Deontology: From the Greek word *deontos*, which means "a duty." Deontology is an approach to ethics that rests on the idea of duty or obligation. It maintains that some actions are good and others are wrong, regardless of their consequences. It is often associated with the idea of rights.

Ecofeminism: A diverse political and philosophical movement whose members share the belief that the oppression of women and the oppression of the natural world are related and that environmental philosophy is deepened if gender issues are integrated into it. In environmental ethics, ecofeminists have tended to adopt approaches based on context and on relationships of care.

Endangered Species Act: A key piece of U.S. legislation passed in 1973 (and subsequently extensively amended). The 1973 act has as its stated aim the preservation of species of fish, wildlife, and plants that "are of aesthetic, ecological, educational, historical, recreational and scientific value to the Nation and its people."

Endemic: In biology, a species is said to be endemic to a particular area if it is found there but not elsewhere.

Environment: Generally, environment means surroundings. It is possible to talk about the economic environment, urban environment, and so on, but most commonly in environmental ethics the term refers to the natural world.

Environmental Ethics: Environmental ethics examines how human beings *should* or *ought to* interact with the environment.

EPA: Acronym for the United States Environmental Protection Agency, founded in 1970.

Gaia: The Greek name for the goddess of earth. Recently, the name was adopted by the scientist James Lovelock to describe his hypothesis that all the constituents of earth, including its living elements, oceans, atmosphere, and rocks, are part of a self-regulating living system that keeps the earth inhabitable for life.

Genetic Engineering: The human technique of manipulating genes (units of deoxyribonucleic acid or DNA) at the level of the cell or the molecule. The expression may be used as a general term to include other similar processes, in particular recombinant DNA technology.

Global Warming: The increase in temperature at the earth's surface caused by the insulating nature of the atmosphere. Nowadays, the term is frequently used to refer to the enhanced warming effect caused by changes in the composition of the atmosphere due to anthropogenic emissions of so-called greenhouse gases.

Greenhouse Gas: Gases believed to be implicated in global warming. The most important greenhouse gases are thought to be carbon dioxide, chlorofluorocarbons, methane, nitrous oxide, and ozone.

Herbicides: Substances used to kill or control unwanted plants.

Hierarchy: A systematically unequal distribution of power or value in which one individual or group is ranked more highly than another.

Individualist: Although this word can be used in different ways, it is used here to describe the belief that an individual always has priority over a collective in ethical decision making. In environmental ethics, this term encompasses ethical positions where the well-being of individual organisms is accorded priority over the health of ecosystems or species. (Indeed, individualists may argue that ecosystems and species are not the kinds of things that can be healthy or not.)

Instrumental Value: Value ascribed to something or to some state of affairs because it is useful—because it is an "instrument" to achieving another end. It thus contrasts with *intrinsic* value.

Intrinsic Value: Although this expression can be used in different senses, it is used in this book to mean value that is an end in itself, value that is not useful for any purpose beyond itself. It thus contrasts with *instrumental* value.

ISEE: Acronym for the International Society of Environmental Ethics.

Land Ethic: This expression is usually used to describe the ethic of the forester and philosopher Aldo Leopold. His most famous statement of what he called the land ethic appeared in *A Sand County Almanac* (1949): "A thing is right when it tends to preserve the stability, integrity, and beauty of the biotic community. It is wrong as it tends otherwise."

Pesticides: Substances used to kill or control unwanted pests such as insects and fungi.

Pollution: To pollute is, literally, to make foul or dirty. In the environmental field, pollution refers to the anthropogenic emission of substances or energy that damage some aspect of the environment or harm living organisms.

Precautionary Principle: The precautionary principle can be defined in many different ways, but generally it states that where there is a threat of serious harm to the environment, lack of scientific certainty should not prevent action being taken with respect to it. Versions of this principle have been widely used in international policy making, including in the agreements at the 1992 Rio Earth Summit.

Rio Earth Summit: Common name for the United Nations Conference on Environment and Development (UNCED), held in Rio de Janeiro, Brazil, in June 1992.

Social Ecology: A philosophical and political movement associated with the work of Murray Bookchin. Drawing on the work of Marx and Engels, and on a range of anarchist writers, social ecologists argue that environmental problems are the result of oppressive hierarchical relationships in human society. Changes in human society, social ecologists argue, will also lead to changes in the way the environment is understood and treated.

Stratospheric Ozone Depletion: The depletion of ozone in the upper atmosphere (stratosphere) caused by the anthropogenic emission of various substances including CFCs and halon gases. The depletion of ozone allows higher levels of harmful ultraviolet radiation to reach the earth's surface, causing damage to human beings and other living things.

Sustainable Development: Defined in 1987 by the World Commission on Environment and Development (WCED) as "development that meets the needs of the present without compromising the ability of future generations to meet their own needs." This is probably the most frequently cited definition, but the term is subject to heated debate in both academic and political circles.

Transboundary: Extending or passing across a boundary. In the environmental field, the term is most often used to refer to the transportation of waste or pollution across national frontiers.

Utilitarian: The most popular consequentialist approach to ethics. It is usually associated with the work of the nineteenth-century philosophers Jeremy Bentham and John Stuart Mill. Mill defined utilitarianism as the "creed which...holds that actions are right in proportion as they tend to promote happiness, wrong as they tend to promote the reverse of happiness." A utilitarian approach to justifying animal liberation has been developed by the philosopher Peter Singer.

Index

Clare Palmer is a lecturer in environmental sciences at the University of Greenwich (U.K.) and an Associate Fellow of the Centre for Environment, Ethics and Society, Mansfield College, Oxford. She has written extensively in the field of environmental ethics and is coeditor of *The Earth Beneath: A Critical Guide to Green Theology* (1992).